MOBILE AD HOC NETWORKS

Bio-Inspired Quality of Service Aware Routing Protocols

MOBILE AD HOC NETWORKS

Bio-Inspired Quality of Service Aware Routing Protocols

G Ram Mohana Reddy
Kiran M

CRC Press
Taylor & Francis Group
Boca Raton London New York

CRC Press is an imprint of the
Taylor & Francis Group, an **informa** business

CRC Press
Taylor & Francis Group
6000 Broken Sound Parkway NW, Suite 300
Boca Raton, FL 33487-2742

Printed on acid-free paper
Version Date: 20160614

International Standard Book Number-13: 978-1-4987-4685-4 (Hardback)

Visit the Taylor & Francis Web site at
http://www.taylorandfrancis.com

and the CRC Press Web site at
http://www.crcpress.com

Printed and bound in the United States of America by Publishers Graphics,
LLC on sustainably sourced paper.

To my parents, B. Manjappa and Padma Manjappa, my wife, Dr. Karishma Kiran, and my daughter, Manasvi Kiran.

— Kiran M.

To my parents, the late Sri G. Ramasubba Reddy and the late Smt. G. Ramakka, and my elder brother, the late Sri G. Ramachandra Reddy.

— G.R.M. Reddy

Contents

List of Figures

List of Tables

Preface

The distinguishing characteristics of mobile ad hoc networks (MANETs) such as increased connectivity, decentralized communication, infrastructure-less environment, easy deployment and maintenance have made MANETs a promising and popular network in today's telecommunication network. These distinguishing features of MANETs are well suited for military applications and disaster applications. Further, MANETs are also used in emergency operations such as search and rescue. Thus, this applicability of MANETs has further influenced its increased popularity among the other networks. Hence, MANETs have become research topics of high interest in research fields. On the other hand, MANETs suffer from the following shortcomings:

1. Due to multi-hop communication models, MANETs suffer from low throughput, high end-to-end delay and increased control packet overhead.
2. Due to the dynamic topology of MANETs, finding a reliable path between the source and the destination node is a challenging task.
3. As MANETs contain resource constrained nodes and channels are shared among the available nodes in the network, providing QoS to the application end users is difficult.
4. Since available bandwidth is smaller, providing quality of service (QoS) in MANETs is more difficult.
5. As communication channels are shared, there are some security issues in MANETs, especially in military applications.

Thus, researchers are working toward finding new and novel solutions to the weaknesses of MANETs. In recent years, a lot of work has been done in an effort to incorporate swarm intelligence (SI) techniques in building adaptive routing protocols for MANETs. As centralized approaches for routing in MANETs lack scalability and fault tolerance, SI techniques provide natural solutions through distributed approaches to the adaptive routing for MANETs. Further, the cross layer desing (CLD), a promising technique used for providing QoS, has not been explored much in developing SI-based routing protocols for increasing the efficiency of mobile wireless networks. One can find very limited work which combines CLD and SI techniques in developing adaptive routing protocols for MANETs.

Thus, this text concentrates on two different techniques, namely SI and CLD, for providing QoS in MANETs. Further, the authors focus on how security could be provided to the application end users through SI principles for MANETs. At the end of the day, reader will know about the salient features of animal societies and insect societies that have influenced researchers in developing adaptive routing protocols for MANETs and exploring the analogy between MANETs working principles and animal and insect societies. Further, the future directions will motivate the researchers to think in the open research directions in the fields of SI and CLD for providing QoS for MANETs.

Acknowledgments

I thank my parents, B. Manjappa and Padma Manjappa, for their constant encouragement. I attribute the completion of this work to the help of my wife Dr. Karishma Kiran, and my daughter, Manasvi Kiran. Without them, this text surely would not have been made possible. I also thank C.G.M Vishwas and Mamatha for their continuous support during the preparation of this text. Finally, I thank my institution, Jawaharlal Nehru National College of Engineering at Shivamogga, Karnataka, India.

M. Kiran

I express my sincere gratitude to the Directors of National Institute of Technology Karnataka Surathkal, Mangalore, Karnataka, India namely Prof. Sandeep Sancheti (during November 2006 to July 2012) and Prof. Swapan Bhattacharya (during August 2012 to December 2015) for their enthusiasm and support towards research in this area of Mobile Ad Hoc Networks and also for writing this book. I also thank my wife, Vijayalakshmi, for her patience and encouragement during the preparation of this book.

G.R.M. Reddy

1

Introduction

CONTENTS

Computer networks can be broadly classified as wired and wireless networks. Wireless networks can be further classified into single-hop and multi-hop wireless networks. Single-hop wireless networks can be infrastructure based (e.g., cellular networks) or infrastructure-less (e.g., Bluetooth). Multi-hop wireless networks can also be infrastructure based (e.g. Wireless Mesh Networks WMNs) or infrastructure-less (e.g., mobile ad hoc networks (MANETs), or wireless sensor networks (WSNs)). Infrastructure-based networks have a fixed infrastructure of network elements such as routers, gateways, access points and base stations. Infrastructure-less networks do not have any fixed

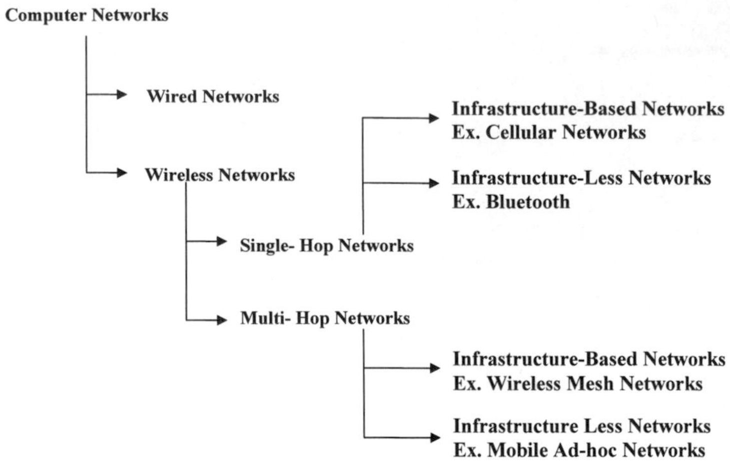

FIGURE 1.1
General Classification of Computer Networks; Ex=Example.

infrastructure and hence nodes act as routers or gateways when required. Figure 1.1 shows the general classification of computer networks.

1.1 Multi-Hop Networks

A wireless multi-hop network is defined by a group of wireless nodes which are dynamically interconnected, forming a network without an infrastructure. Communication in this type of network occurs in a decentralized way. Therefore, the network nodes rely on each other for communication and there will be one or more intermediate nodes that receive and forward the packets along the communication path from the source node to the destination node. Thus, each node is acting like a router.

Multi-hop wireless networks are cost effective since they avoid wide deployment of cables. Compared to single-hop networks, they can extend the coverage area of a network and thus improve connectivity. They also save energy and transmission power required over long links by using multiple short links. Multi-hop networks use wireless media more efficiently, thereby yielding higher data rates with high throughput [24]. Conversely, they suffer from routing complexity, need for path management, and delays due to multi-hop relaying and security threats. Based on their infrastructures, these networks can be classified as infrastructure-based [e.g.,wireless mesh networks) and infrastructure-less (e.g.,mobile ad hoc networks).

1.1.1 Mesh Networks

A wireless mesh network (WMN) is dynamically self-organizing and self-configuring and its nodes will have mesh connectivity. Each node in the WMN plays a dual role, i.e., it acts as a host, as well as a router, forwarding packets on behalf of other nodes. The nodes in the mesh networks are classified into two types: mesh routers and mesh clients. The mesh routers, other than the basic router capabilities, contain additional routing functionalities such as bridges, gateways, higher transmit power, multiple receive and transmit interfaces and unlimited power supply, to support mesh networking. The mesh router is the backbone of WMN and it has minimal mobility and is responsible for connecting WMNs with other networks. WMNs are undergoing rapid progress and finding various practical uses. Traditional nodes with wireless network interface cards (NICs) can be connected directly to wireless mesh routers whereas the nodes without wireless NICs can connect to WMNs through Ethernet [8].

Whenever a mesh client wants to send a message to another mesh client (not in its cell) it first sends the message to its mesh router. From the mesh router the packets are forwarded over the backbone of the mesh through multiple hops until they reach the mesh router (or the gateway) which can forward the packets to the destination node immediately. Figure 1.2 depicts the architecture of a wireless mesh network in which a mesh router is connected to the wired network.

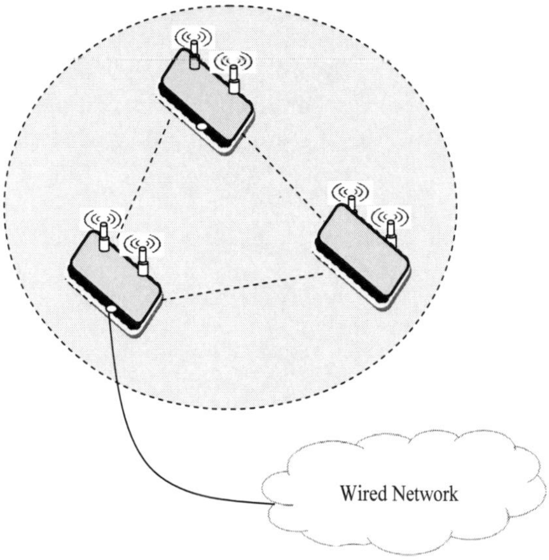

FIGURE 1.2
Mesh Network Architecture

1.1.2 Mobile Ad Hoc networks

Mobile ad hoc networks (MANETs) belong to the family of multi-hop wireless networks in which a group of mobile nodes are dynamically interconnected and cooperate with each other to maintain network connectivity. Ad hoc networks have no fixed network infrastructure or administrative support whereas a conventional wireless network requires some form of the fixed network infrastructure and centralized administration for its operation. Ad hoc networks dynamically create a wireless network among themselves any time, anywhere and exhibit as self-creating, self-organizing and self-administrating behaviors. The physical characteristics, organizational format and dynamic topology are the distinguishable parameters of MANETs.

These types of networks are useful in any situation where temporary network connection is needed. Due to the limited transmission range of wireless network interfaces, multiple network hops may be required to transmit the data between the source and the destination nodes across the network. A MANET network poses a significant technical challenge because of the many constraints imposed by the underlying network [20] [23] [11].

Quality of service (QoS) in MANETs in terms of bandwidth, delay, delay jitter, and packet delivery ratio is not matter-of-fact because of node mobility and the shared nature of the wireless medium. With the user's desire for real-time applications, new challenges in developing new protocols for MANETs have opened. Among those challenges, two interesting and important needs are to support QoS and provide security. Two categories of QoS are proposed: soft QoS and QoS adaptation. A soft QoS feature is that failure to meet the requirement is affordable. Certain applications can optimize their performance based on the feedback about the availability of the resources for QoS adaptation.

1.2 History of Mobile Ad Hoc Networks

Ad hoc networking finds its roots in 500 BC when Darius I, the king of Persia, found it difficult to communicate messages from the capital to outlying areas of his kingdom. The king wanted to build a communication system that was effective and fast enough to reach the outlying areas. Darius used groups of shouting men standing on tall towers to pass messages from the Persian capital to the outlying areas of the kingdom in an ad hoc manner. The king found the new communicating system fast and effective.

The credit of inventing ad hoc communication system also goes to the ancient and tribal societies who used to communicate among themselves through drums, trumpets or horns. Motivated by the need to provide network access and communication to mobile hosts and terminals, the U.S. Defense Advanced Research Projects Agency (DARPA) invented the Packet Radio Network (PRNET) in 1973. It marks the first attempt to invent a distributed multi-hop wireless network for mobile nodes. PRNET exhibited several features. PRNET operates when the mobile nodes are in motion; installation and deployment were much simpler and quicker and finally, reconfiguration and redeployment were also simpler. PRNET also exhibited features such as use of single channel and easier channel management, self configuring based on the situation and dynamic routing. The success of PRNET was followed by Survivable Radio Networks (SURAN) in the 1980s from DARPA which was anticipated to provide ad hoc networking with small, low cost and low power devices. In the 1980s the Mobile Ad Hoc Networking Working Group, a subdivision the of Internet Engineering Task Force (IETF) was formed to develop standards for the protocols and functional specifications for ad hoc networks.

Inspired by multi-hop wireless working scenario, in 1994 a one-hop wireless communication infrastructure referred to as Bluetooth was invented by Ericsson, a Swedish communication equipment maker. Bluetooth works on heterogeneous devices. A single-hop point-to-point wireless infrastructure or wireless device can exchange voice or data among each other.

Alternatives to infrastructure-less multi-hop wireless networks, recent advancements include infrastructure-based multi-hop wireless networks such as multi-hop cellular networks (MCNs) and self-organizing packet radio networks with overlays (SOPRANOs) that combine the features of both cellular and ad hoc networks. The performance of these hybrid networks is impressive and encouraging [3].

1.2.1 Differences between Cellular Networks and MANETs

In a cellular network, also known as an infrastructure-based network, the call setup between two nodes starts at the base station in a single-hop scenario. Figure 1.3 depicts the working principle. If node A wants to set up a path to node B, the communication link is established through the nearest reachable base station (BS1). The call setup and maintenance decisions are made by the base station; thus routing decisions are centralized. This system simplifies routing and resource management issues for a cellular network. After a call is set up, the end user has fairly seamless connectivity with few call breakups and packet drops. Since the base station controls the network, ensuring QoS is much easier. Providing security to the applications and protocols is also simpler because decisions are made at a single point (the base station). However, the time and cost needed to deploy a cellular network is higher than the costs for MANETs.

In MANETs, the base station concept is removed to suit the applications intended and hence MANETs are infrastructure-less networks. For transmitting voice and data, MANETs depend on cooperative communication where each node acts as a host as well as a router. Figure 1.4 shows the working principle of MANETs; if node A wants to communicate with node B, the call is set up through the nearest neighbors available in a multi-hop scenario. As shown in Figure 1.4, the source node A establishes call to the destination node

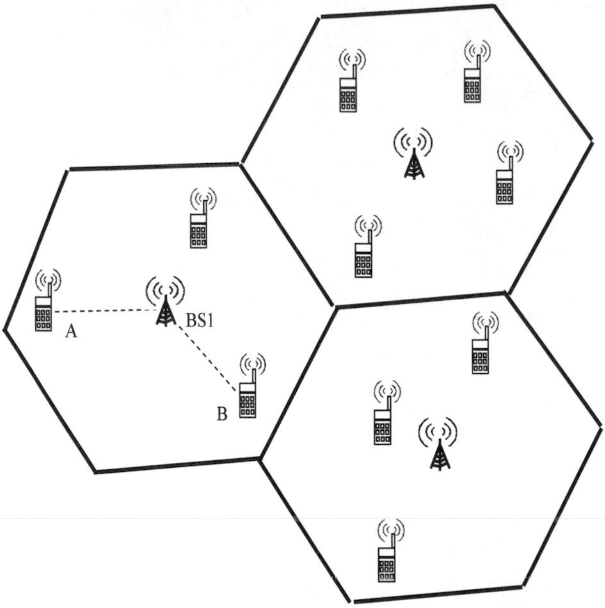

FIGURE 1.3
Cellular Network Working Scenario.

B through intermediate nodes X, Y and Z. Then, on behalf of node A, the nodes X,Y and Z receive and transmit the data to the destination node B.

Since there is no base station, routing and resource management in MANETs is carried out by individual nodes only. To meet the dynamic environment scenario, MANETs employ distributed routing to achieve an effective routing. Due to the mobility of nodes, MANETs suffer from the frequent path breakups and quick reconfiguration of broken paths is difficult in MANETs; hence it requires a more sophisticated routing protocol to handle distinguishing features of MANETs. Further, providing QoS and security in MANETs is difficult and challenging due to the resource constrained nodes and node mobility. On the other hand, the time and cost incurred for deployment of MANETs is very low as compared to cellular networks. Thus MANETs find applications in several areas due efficient and cost effective deployment features; some of the applications are discussed below:

1.2.2 Applications of Mobile Ad Hoc Networks

1. Military applications: The primary need for MANETs is military application since deploying a cellular network in the field or in the enemy's territory is impossible. Setting up an ad hoc network among the soldiers and military facilities is quick and feasible.

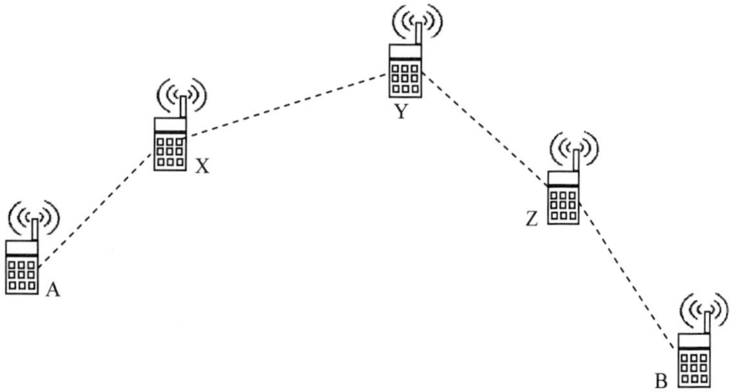

FIGURE 1.4
MANET Working Scenario

2. Disaster applications: In case of natural calamities such as earthquakes or tsunamis where a base station may not work properly, MANETs easily establish the communication. Among the available mobile devices, an ad hoc infrastructure could be set up and messages could be sent to a rescue team for help. As deployment time of an ad hoc network is minimal, a quick network setup in such situations without any delay and a rescue could be achieved as quickly as possible.
3. Distributed file sharing: A quick communication infrastructure could be developed for a meeting or conference using MANETs without the need of wires, cables and other networking devices for sharing files and other resources among the participants. The established network infrastructure is temporal and will exist only for the duration of the meeting or conference.

We can also find several other applications of MANETs such as public safety by providing access to the remote regions and areas where deployment of cables is not possible (Murthy and Manoj 2005).

1.3 Routing in Mobile Ad Hoc Networks

MANETs belong to the family of multi-hop wireless networks in which a group of mobile nodes cooperate with each other to maintain network connectivity. MANETs have no fixed network infrastructure or administrative support but have valuable features such as self-creation, self-organization and self-administration. However, the network communication in MANETs is decentralized due to the limited transmission range of mobile nodes and thus multiple hops are required to transmit the data between the source and destination nodes. In MANETs, a path between the source and destination nodes contains one or more intermediate or relay nodes which receive and forward the packets, and thus each node acts as router. MANETs extend the coverage area of the network without additional infrastructure, thereby providing better connectivity with less energy and transmission power. However, they suffer from routing complexity, path management overhead and extra delay due to multi-hop relaying. Further, providing QoS is a critical and challenging task in MANETs due to their dynamic network topology [17].

1.3.1 Design Challenges for Routing Protocols

The performance of MANETs depends on the routing technique and also on the routing metrics such as shortest path and bandwidth, delay. Since MANETs pose a significant technical challenge because of constraints imposed by the underlying network, routing is the most critical and challenging task; the main challenges in routing are highlighted below [9] [7] [4].

1. QoS aware routing is important since MANETs are expected to provide enhanced end-to-end services such as high throughput, low end-to-end delay, less packet drops, and minimal control packet overhead. under dynamic network topological conditions.
2. Scalability is vital in MANETs routing and max degrade the throughput with increased network control packet overhead under a high node density scenario. Thus a MANET routing protocol should be scalable to higher node density in such a manner that the control packet overhead is minimized and maximum throughput is achieved with less delay.
3. Energy efficient routing is critical for maximizing the network lifetime of MANETs since mobile nodes (mobile phones, lap tops and personal digital assistants (PDAs)) have limited energy. Nodes with poor energy levels may break the communication link or cause partition in the network in the worst case scenario.
4. Route maintenance and recovery should be important characteristics of routing protocols in MANETs since the hostile environment can impact or damage the available nodes leading to breaking up the path and affecting the overall throughput.
5. Providing security is one of the important challenges routing since MANETs are vulnerable to attacks such as confidentiality, privacy, authorization, and authentication.
6. Handling hidden and exposed terminal problems during routing is

another challenging task in MANETs since they may be misinterpreted as link breakages or non-reachable nodes.

7. Most of the MANETs routing protocols lack load balancing and hence load balancing across multiple paths is very important to increase efficiency.

Present MANETs provide all kinds of multimedia traffic for users and the need for novel QoS aware routing protocols for MANETs present new challenges; a brief overview of QoS is given in the next section.

1.4 Quality of Service (QoS) Issues

QoS is associated with the performance of computer communication networks and according to ITU-T standard E.800, QoS can be defined as the collective effort of service performance which determines the degree of satisfaction of a user of a service. Further, IETF contributed many protocols (Intserv, Diffserv, RSVP and MPLS) to introduce QoS to the Internet; and also discussed QoS routing and established the IETF QoS Routing Working Group [12] in order to provide QoS to the end users.

Since the current generation of MANETs supports all kinds of multimedia traffic, it requires QoS to cover bandwidths, end-to-end delays, jitters, latency, and other issues. For that reason QoS has become a vital consideration for designers of routing protocols for MANETs. QoS should cover the various layers of the protocol stack, from the physical layer to the application layer. Each layer is responsible for providing QoS in different ways. Table 1.1 lists examples of QoS at various layers [23].

Ensuring service quality in the network layer involves developing a state-dependent QoS aware routing protocol that searches for optimal routes with sufficient network resources to handle bandwidth, end-to-end-delay, jitter, and latency issues. The network environments of MANETs fluctuate and network resources are shared among different users. Thus routing represents a QoS challenge under such dynamic network conditions. For this research, current research focuses on cross layer design (CLD) and swarm intelligence (SI) for designing and developing QoS aware routing protocols for MANETs [6].

1.5 Security Issues in MANETs

Earlier research on MANETs focused on solving primary problems that arose from specific characteristics such as dynamic topology, multi-hop relaying,

TABLE 1.1
QoS at Different Layers

Layer	**QoS**
Application	Simple and flexible user interface
Transport	Reliable end-to-end packet delivery
Network	Throughput, end-to-end delay
MAC	Variable bit error rate
Physical	Good transmission quality

and routing table maintenance. The solutions relied on neighboring nodes for proper functioning and security was neglected. A proper solution should be secure and tamper resistant.

Recent studies concentrated on algorithms and models designed to protect and thus ensure proper functioning of MANETs protocols and applications. In addition to the inherent vulnerabilities of MANETs, some of the underlying network features also contribute to security breaches, for example, reliance on radio communications, unreliable wireless channels, and bandwidth limitations [5].

Ad hoc networks have become more common and producers of ad hoc networks cannot neglect security problems, particularly when the networks are used for sensitive military applications. MANETs are more vulnerable to attacks due to shared wireless media, reliance on neighboring nodes for communication, and insufficient infrastructure. Some network susceptibilities to attacks are listed below:

1. Since ad hoc networks are susceptible to wireless channels, eavesdropping of messages and injecting fake messages into the network are easy for an intruder.
2. Since nodes in ad hoc networks are not substantially protected, capturing of nodes and obstructing nodes are clearly Vulnerabilities.
3. The absence of infrastructure makes the nodes in an ad hoc networks communicate by means of neighboring nodes. These neighbouring nodes may be cooperative or uncooperative. Uncooperative nodes may cause ad hoc networks to malfunction and fail to function on the expected lines.

In an ad hoc network at society level, one can expect security requirements such as availability, integrity, confidentiality, privacy, authorization, authentication, non-repudiation and freshness. Security is needed in all layers of a protocol stack, from the physical layer to the link layer and from the network layer to the transport layer. Attack on a physical layer can cause communication channel disruption by means of jamming (redundant noise to disrupt the ongoing signal) and tampering (physical damage to the nodes). On the link layer, attackers will try to deplete the energy of the nodes by creating unnecessary duplicate packets and eating up the link layer time to delay response to the application. At the network layer, an attacker can disrupt ongoing sessions with wrong routing information. Further, flooding of hello packets also disturbs the smooth operation of the protocols. Finally at the transport layer, an intruder will flood the packets and can use desynchronization (sending fake sequence numbers to the source) to disrupt the service. Due to the behavior exhibited by MANETs, the security solutions employed for traditional wireless networks cannot be directly employed.

Many security issues are still unexplored in MANETs and need to be studied and solved [16]. Recent research on security involves finding the loopholes of existing protocols and thereby finding security solutions for the explicit attack model in mind. A specific security model may work well in certain scenarios but may fail during unexpected and unanticipated attacks. Therefore, a detailed study and secured model implementation are required for ensuring the security in all layers and can provide the security for both known and unknown security threats in MANETs. The success of cross layer design (CLD) in other fields and applications has led researchers to use CLD to provide security for MANETs also. Several studies have revealed that CLD is a solid security option for MANETs.

1.6 Cross-Layer Design

Due to the variable nature of the wireless communication environment, the traditional layered architecture does not function efficiently in MANETs. However, network tasks are distributed among different layers and the service hierarchy has been defined for each layer accordingly. Further, the traditional layered architecture does not allow direct communication with nonadjacent layers; thus efforts have been made to improve the performance of the protocol stack by using the cross layer design (CLD), thereby providing stack-wise network interdependence. There are three main reasons for using the CLD approach in developing adaptive routing protocol for MANETs: (1) dynamic nature of wireless links (2) chance of opportunistic communication in the wireless links and (3) the new modalities of communication offered by the wireless medium. In CLD, a layer is not treated as a complete independent functional entity, but the information of each layer can be shared among other layers in the protocol stack [22]. Further, CLD has emerged as an important design option when compared to the monolithic and layered protocol stacks and thus attracting several researchers for developing adaptive QoS aware routing protocols for MANETs.

Figure 1.5 shows the working principle of CLD for a three-layer model (LR1, LR2, and LR3) based on ISO's OSI reference model. LR1 cannot

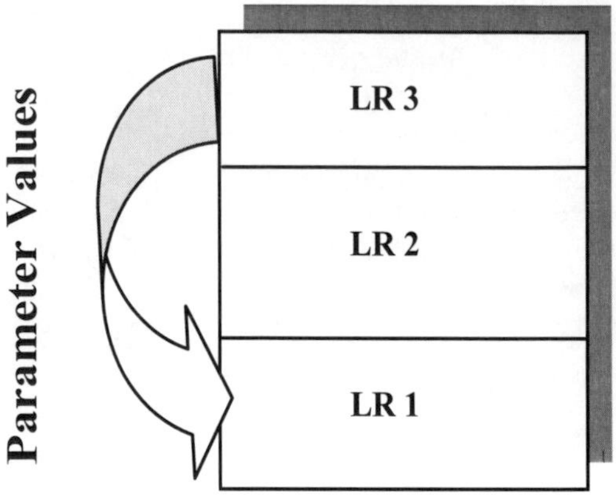

FIGURE 1.5
Example of CLD: Three-Layer Model

communicate with LR3 since it is not adjacent but an interface can be developed in such a manner that LR1 takes LR3 parameter values during runtime.

Some of the examples of the CLD approaches are: (1) the link layer may tune transmit power of the Physical layer to control the bit-error rate; (2) the application layer can get the information about packet loss from the network layer under different situations; (3) the link layer may provide an adaptive error control mechanism based on the TCP retransmission timer information from transport layer [22].

Figure 1.6 shows the different CLD approaches. According to the flow of information across the protocol stack, the newly created interfaces can be categorized as:

- Upward: flow of information is from lower layer to upper layer (Figure 1.6a).
- Downward: flow of information is from upper layer to lower layer (Figure 1.6b).
- Back and forth: flow of information is iterative between two layers (Figure 1.6c).

Layered architecture can also involve merging of adjacent layers (Figure 1.6d), creation of new interfaces (Figure 1.6e), and vertical calibration across layers (Figure 1.6f). MANETs performance may be further enhanced by using efficient techniques to solve the optimization problems to provide the maximum QoS and security to the users. Optimization techniques attempt to find the best solutions among all the feasible solutions. Traditional routing techniques use the first available solution by neglecting the remaining solutions that may be better than the first available solution. Further, traditional routing techniques use the first available solution untill the end of the current session even

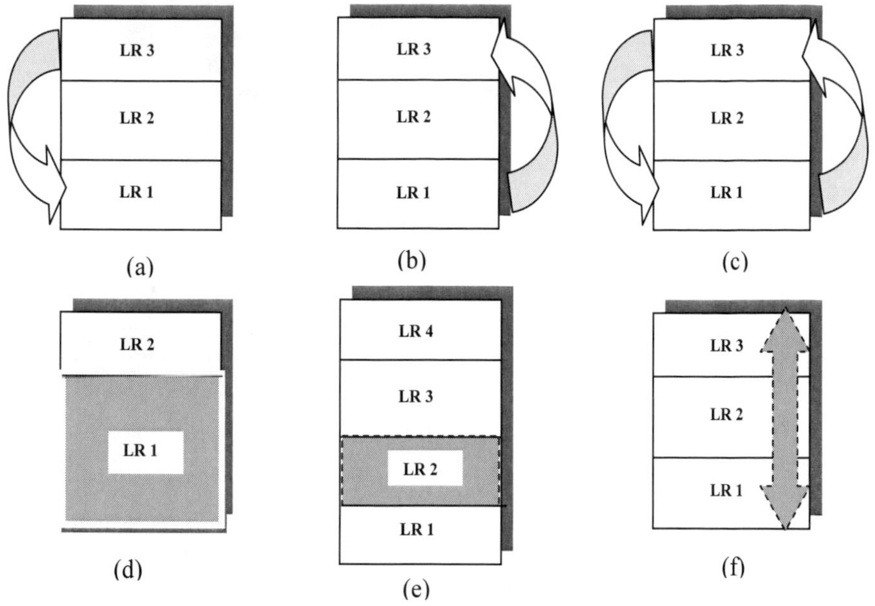

FIGURE 1.6
Different Approaches for CLD

though the selected solution may be optimal or non-optimal. One of the problems with the dynamic networks such as MANETs is that the optimal solution chosen for the current session may become non-optimal and sub-optimal solutions may become optimal over time. Thus an adaptive routing protocol which reflects the current status of the network is very important in MANETs to provide efficient QoS service to users. Bio Inspired computing is one of the popular optimization and adaptive techniques which is grabbing researchers attention to solve optimization problems in telecommunication networks. Bio-inspired computing is explained in detail in the coming section.

1.7 Bio-Inspired Computing

Bio-inspired computing is a popular technique for solving the optimization problem and can be defined as an approach to designing algorithms based on distributed problem-solving inspired by the behavior of social insects and other animal societies [27]. Bio-inspired computing is a technique by which behaviors of swarms of insects or other animals are studied and emulated to solve complex real world problems. The bio-inspired algorithms have several applications in the areas of engineering, logistics and telecommunications in general and in particular the telecommunications domain has achieved remarkable success [15] [21] [25]. Bio-inspired computing techniques can be further classified as swarm intelligence (SI, for example, ant colony optimization (ACO)), termite colony optimization, and non-SI-based techniques such as bat technology, rat technology, and dolphin technology. The next section covers details of SI.

1.7.1 Swarm Intelligence Techniques

SI is inspired by the behaviors exhibited by the social insects (ants, termites), flocks of birds, bees, and schools of fish. An SI system is driven by four primary principles: positive feedback, negative feedback, randomness and multiple interactions [15].

1. Positive feedback is used to emphasize the best solution in the SI system and will make other insects follow the emphasized solution.
2. Negative feedback is used for removing old or stale solutions in an SI system which consequently weakens the probability of choosing these solutions. If a solution is found to be poor or-non optimal, it will be negatively reinforced and and direct other insects not to follow it. Positive and negative feedback should be equally balanced to yield an optimal solution.
3. Randomness: As many solutions will be available in the SI system, an optimal solution may become non-optimal if its quality degrades or whenever a new solution is available. Thus exploring all available solutions randomly is very important to ensure selecting the best available solution.
4. Multiple interaction: The SI system often contains tens to millions of individuals and the information collected by this large population is communicated through multiple interactions.

Recently another principle called stigmergy was added to the SI system [19]. Stigmergy is a method of indirect communication among the members of the same societies in SI systems and it can increase the communication robustness and asynchronous information transfer among the individuals.

Swarms such as ants and termites are distributed in nature and these social insects are simple, independent, interdependent and co-operative in nature. While building a nest or finding the shortest path to the food source, these social insects exhibit wonderful characteristics such as adaptability, scalability and robustness. Pheromones are means of communication among these social insects and are signal carrying chemicals; pheromones are broadly classified into two types: releasers and primers. Upon receiving releaser pheromones, insects exhibit instantaneous behavior (alarm, sex and recruitment or trail pheromones) while primer pheromones cause a delayed effect on insect behavior (sex pheromones by honeybee queen) [18]. Recruitment or trial pheromone act as navigational aids and direct other insects to a distant location e.g., food source varying from hundred meters in bees to meters in insects. These trails are followed by foraging insects and are positively or negatively reinforced based on the quantity and quality of the food source. Trial pheromones are also used to recruit workers for colony emigration.

In social insects, alarm pheromones are the most commonly produced class after sex pheromones and are normally released when there is a threat to the nest or to the other insects. On reception of these releaser pheromones, insects show immediate response by dispersing from the source of alarm a pheromone that detects the danger. Normally the social insects like ants, bees, wasps and termites attack the enemy which threatens the nest or exhibit a panic behavior and escape when the defense is not realistic [26]. This will be illustrated by placing a dead ant near the nest entrance; most of the nearby ants will exhibit a panic behavior.

Mosquitoes, also referred to as gnats, identify their prey based on the chemical and heat omitted from the body of the prey. Mosquitoes have sensors to reveal chemicals (carbon dioxide and lactic acid) which are released during the breathing and sweating of mammals and birds. Further, mosquitoes can also sense the heat of warm-blooded mammals and birds, thereby tracking their prey. Thus the intelligence behavior used by mosquitoes to identify their prey has inspired researchers to design and develop efficient bio-inspired algorithms for MANETs.

Birds travel in flocks of V shape where the spherehead takes the entire burden and reduces the up thrust required by the rest of the birds in the flock; thus the other birds in the flocks save energy. When the spherehead losses its energy, it will be replaced by another bird to take the burden of the flock. No bird in the flock will have any idea about its position in the flock and no specific bird directs the movement of flocks. Thus birds can travel a long distance without losing much energy and this feature inspired many researchers to develop energy efficient bio-inspired routing algorithms for MANETs.

A honey bee colony can intelligently choose the quality and quantity of the food source through simple behavior of individual bees. A honey bee colony can contain several kinds of worker bees such as food storers, scouts and foragers which are responsible for recruiting and foraging for beehive. Bees perform dances (waggle dance and tremble dance) to communicate with their fellow bees.

A fish school consists of several individuals of the same age and size that synchronously swim at the same speed and direction usually to avoid predators. This synchronous swimming also helps the fish in the population to seek food and mates and optimize environmental parameters (e.g., temperature).

Thus the characteristics (adaptability, scalability and robustness) and intelligent behavior of these social insects (optimal path finding by ants and the hill building nature of termites), mosquitoes (sensing capability to track prey), flocks of birds (energy saving concept of the flock), bee colony (technique of conveying the quality and quantity of the food to other bees in the colony) and fish schools (synchronous swimming to avoid predators) may be exploited for design and development of routing protocols for MANETs [2] [14] [10].

1.7.2 Non-Swarm Intelligence Techniques

In addition to Swarms, (ants, bees, flocks of birds and termites) there exist other animal societies such as mammal bats that have also motivated researchers for designing adaptive bio-inspired algorithms for solving complex real world problems. The fascinating echolocation characteristic of mammal bats allows them to find their prey (distance and movement direction) in complete darkness. The bats produce very loud pulses and listen for the echo that bounces back after hitting the prey and the surrounding objects; depending on the species and obstacles, the pulse varies when it echoes back. According to a study, micro bats species build a three-dimensional scenario of surrounding environment using the time delay from the emission of pulse to the detection of an echo and the loudness variations of the echoes. Thus, bats can detect the distance and orientation of the target, type of prey and its moving speed using the Doppler effect [1].

Dolphins, like bats, also use echolocation to hunt their prey. Dolphins direct click towards prey and listen to the strength of the echoed clicks after hitting the prey. The echolocation enables dolphins to find the distance, size, shape and movement direction of prey.

Rats have the incredible characteristic of learning from situations they face. If a rat becomes aware of an unfavorable or dangerous situation, it can take the necessary steps to avoid it without disturbing its day-to-day activities.

Non-swarm intelligence (Non-SI) techniques have not been explored as thoroughly as SI techniques for solving real-world problems, particularly in the telecommunications field. The next section discusses the motivation that led to the writing of this text.

1.8 Motivation

Existing layered protocol stacks do not function efficiently in mobile wireless environments due to the limited and highly variable nature of mobile devices [28]. Advances in layered architecture should improve the performance of wireless networks significantly.

Current research has focused on CLDs in designing and developing novel routing algorithms for MANETs to ensure QoS for the applications. Centralized approaches for routing in MANETs have failed to provide scalability and fault tolerance; SI techniques provide natural solutions through distributed approaches to adaptive routing in MANETs.

The characteristics and behaviors of social insects could be adapted in designing and developing adaptive QoS routing algorithms for MANETs [2] [14]. The mobile nodes found in MANETs are capable of monitoring network status and data processing. MANETs can be made context aware with the help of the local monitoring capabilities of mobile nodes.

Traffic carried by MANETs is expected to carry mixes of real-time multimedia, nonreal-time file transfers, and other functions. Providing QoS for these applications in MANETs is difficult because of their varying characteristics. Designing novel bio-inspired QoS aware routing protocols for MANETs is attracting more attention because traditional routing protocols failed to meet most network service requirement. Security was neglected in earlier research but is receiving ample attention now because of the unique security needs of MANETs protocols.

These developments motivated us to write a detailed study of current trends in applying CLD and bio-inspired computing to MANETs to ensure adequate security. The possible application of behaviors of social insects and animal societies to nodes in MANETs is also explored in this book.

1.9 Organization of the Book

In Chapter 2, the characteristics and behaviors of various swarms are discussed. The communications patterns used by several species to find food, return to their nests, or avoid danger are discussed in detail. This book concentrates on behavioral studies of ants, termites, bees, birds, wolves, elephants, and other animal societies. Chapter 2 concludes with an analogy between SIand MANETs routing techniques.

Chapter 3 discusses and analyzes state-of-the-art bio-inspired routing techniques in MANETs. It highlights the working principles of legacy routing protocols, then covers the applications of SI and non-Sl solutions to routing problems in MANETs. A taxonomy based on the code of practice, required properties, and shortcomings of the relevant algorithms is included. The chapter concludes with a discussion of hybrid routing protocols in which the behaviors of multiple species are combined to develop solutions for MANETs routing problems.

Chapter 4 covers QoS for real-time applications in MANETs and parameters that can be used in different layers of the protocol stack that will satisfy the QoS needs of the applications. Also covered are SI solutions to QoS and classification of SI-based QoS protocols in MANETs. The categories are classified as Sl-based QoS aware protocols using CLD and Sl-based QoS aware protocols not using CLD.

MANETS security issues are discussed in Chapter 5. The chapter starts by detailing MANETs vulnerabilities and the types and impacts of attacks, then explains security threats faced by each layer of the protocol stack. The final sections deal with SI solutions to security problems and the taxonomies of SI-based secured routing protocols for MANETs.

Chapter 6 presents conclusions and predicts future directions and challenges for researchers.

1.10 Summary

The absence of wires and cables allowed mobile operation of ad hoc networks, thus providing users with portability and flexibility. The lack of wires and cables also made deployment, addition, and deletion of nodes in ad hoc networks much easier than performing those functions in wired networks.

Research on applications of SI in routing for telecommunication networks has gained wide acceptance. Many SI-based adaptive routing algorithms have been developed for MANETs. Centralized approaches to routing in MANETs lack scalability and fault tolerance. SI techniques provide natural solutions through distributed approaches to adaptive routing.

The inability to communicate of non-adjacent layers in traditional protocol stacks in MANETs was a serious drawback because it caused degraded performance and inefficient use of available resources. Current research focuses on allowing non-adjacent layers to communicate with each other to achieve better performance and resource allocation.

This book focuses on two important aspects of MANETs: QoS and security. It explains how traditional techniques fail to provide effective QoS and security to end users. It explains SI- and CLD-based solutions for MANETs. It also details comprehensive studies and taxonomies of SI-based QoS aware, SI-based QoS aware with CLD, SI-based secured routing, and SI-based secured routing using CLD protocols.

Glossary

Bio-Inspired Algorithms: Algorithms based on the working principles of social insects or other animal societies.

Swarm: A group of animal, insect or other species serving some purpose.

Pheromone: A chemical substance used by insects to communicate among other insects of same kind that evaporates over time.

Echolocation: A signal (pulse in the case of bats, and ticks in the case of dolphins) is produced and bounced back to the source after hitting the objects. Echolocation features are found in mammals such as, bats and dolphins, that use this technique to find their prey, its distance, movement direction and size.

Exercises

Part A Questions

1. Which of the following options are single-hop infrastructure-based wireless networks?
   (a) Bluetooth
   (b) Cellular
   (c) WMN
   (d) MANET
2. The network that combines with cellular and ad hoc networks is
   (a) PRNET
   (b) MCNS
   (c) SOPRANO
   (d) None of the above
3. PRNET is
   (a) Packet radio network
   (b) Packet route network
   (c) Packet remote network
   (d) None of the above
4. In MANETS each node in the network behaves as a
   (a) Bridge
   (b) Router
   (c) Gateway
   (d) Firewall
5. Multi-hop relaying causes MANETs to suffer
   (a) Routing complexity
   (b) Path management
   (c) Delay
   (d) All of the above

6. MANETs are expected to provide enhanced end-to-end services such as
   (a) High throughput
   (b) Low end-to-end delay
   (c) Less packet drops
   (d) All of the above
7. A problem in MANETs routing that degrades the throughput is
   (a) Reliability
   (b) Integrity
   (c) Scalability
   (d) None of the above
8. QOS is valuable in MANETs for designing
   (a) Routing protocol
   (b) Load balancing
   (c) Terminal problem
   (d) Route maintenance
9. Communication in a multi-hop wireless network will be
   (a) Decentralized
   (b) Centralized
   (c) Predefined
   (d) None of the above
10. Stigmergy communication among members of the same societies in SI systems is
    (a) Direct
    (b) Indirect
    (c) Both (a) and (b)
    (d) None of the above

Part B Questions

1. Explain cross layer design.
2. Discuss the security issues in MANETs.
3. Differentiate swarm Intelligence techniques and non-swarm intelligence techniques.
4. Describe cellular networks and MANETs.

Part A Answers

1. b
2. c
3. a
4. b
5. d
6. d
7. c
8. a
9. a
10. b

References

[1] M. Airas. Echolocation in bats. In *Proceedings of Spatial Sound Perception and Reproduction*, 2003.

[2] G. K. C. Kolias and M. Maragoudakis. Swarm intelligence in intrusion detection: a survey. *Computers and Security*, 30(8):625–642, 2011.

[3] B.S. Manoj and C. Siva Ram Murthy. *Ad Hoc Wireless Networks.* Pearson, 2011.

[4] L. Chen and W. B. Heinzelman. A survey of routing protocols that support QoS in mobile ad hoc networks. *Networks*, 21(6):10–16, 2007.

[5] Lyes Khelladi, Djamel Djenouri and Nadjib Badache. A survey of security issues in mobile ad hoc and sensor networks. *Communications Surveys*, 7(4):2–28, 2005.

[6] F. D. Gianni Di Caro and L. M. Gambardella. Swarm intelligence for routing in mobile ad hoc networks. In *IEEE Swarm Intelligence Symposium*, pages 76–83, 2005.

[7] L. Hanzo and Tafazolli. A survey of QoS aware routing solutions for mobile ad hoc networks. *Communications Surveys Tutorials*, 9(2):10–16, 2007.

[8] Weilin Wang, Ian F. Akyildiz, Xudong Wang. Wireless mesh networks: A survey. *Journal of Computer Networks*, 2005.

[9] M. C. Imrich Chlamtac and J J.N.Liu. Mobile ad hoc networking: imperatives and challenges. *Ad Hoc Networks*, 1:10–16, 2003.

[10] Jorge J. Ruiz-Vanoye et al. Meta-heuristics algorithms based on the grouping of animals by social behaviour for the travelling sales problem. *Internatinal Journal of Combinatorial Optimization Problems and Informatics*, 3(3):104 –123, 2012.

[11] David B. Johnson, Yih-Chun Hu, Jorjeta Jetcheva, Josh Broth, David A. Maltz. A performance comparision of multi-hop wireless ad-hoc network routing protocols. In *MOBICOM 98, Dallas.* ACM, 1998.

[12] X. Masip-Bruin, et al. Research Challenges in Qos Routing . *Computer Communications*, 29(5):10–16, 2006.

[13] J. L. Mohapatra P. and C. Gui. QOS in mobile ad hoc networks . *Wireless Communications*, 10(3):10–16, 2003.

[14] G. A. D. C. Muhammad Saleem and M. Farooq. Swarm intelligence based routing protocol for wireless senson networks: survey and future directions. *Information Sciences*, 181(20):4597–4624, 2011.

[15] F. Neumann and C. Witt. Bioinspired computation in combinatorial optimization algorithms and their computational complexity. *Natural Computing Series*, 2010.

[16] N.V. Verde R. Di Pietro, S. Guarino and J. Domingo-Ferrer. Security in wireless ad-hoc networks: a survey. *Journal of Computer Communications*, 51:1–20, 2014.

[17] R. Ramanathan and J Redi. A brief overview of ad hoc networks: challenges and directions. *IEEE Communications Magazine*, 40(5), 2022, 2002.

[18] F. E. Regnier and J. N. Insect pheromones. *Journal of Lipid Research*, 14, 1968.

[19] M. Roth and S. Wicker. Termite: ad-hoc networking with stigmergy. In *IEEE Global Communication Conference*, 5, pages 2937–2941, 2003.

[20] Amitabh Mishra and Satyabrata Chakrabarti. QOS issues in ad-hoc wireless networks. *IEEE Communications Magazine*, 201–265, 2001.

[21] G. Singh et al. Ant colony algorithms in MANETs: a review. *Journal of Network and Computer Applications*, 35(6):1964–1972, 2012.

[22] V. Srivastava and M. Motani. Cross layer design: a survey and road ahead. *IEEE Communication Magazine*, 43(12):112–119, 2005.

[23] C. S. Raghavendra, Suesh Singh, Mke Woo. Power-aware routing in mobile ad-hoc networks. In *MOBICOM 98, Dallas.* ACM, 1998.

[24] Maria Kihl, Veselin Rakocevic, Vasilios Siris, Torsten Braun, Andreas Kassler and Geert Heijenk. Multi-hop wireless network. In *Traffic and QoS Management in Wireless Multimedia Networks, Lecture Notes in Electrical Engineering*, volume 31, pages 201–265, 2009.

[25] K. Vivekanand, Jha Khetarpal and M. Sharma. A survey of nature-inspired routing algorithms for MANETs. In *3rd IEEE International Conference on Electronics Computer Technology*. volume 6, pages 16–24, 2011.

[26] T. D. Wyatt. *Pheromones and Animal Behaviour: Communication by Smell and Taste*. Cambridge University Press, 2003.

[27] E. Bonabeau 1999.

[28] Raisinghanie et al. 2002

2

Nature-Inspired Routing Algorithms for MANETs

CONTENTS

In the telecommunications field, finding an optimal path between a source node and destination node is very important for ensuring certain aspects of path quality such as high bandwidth and fewer end-to-end delays for end users. A given source and destination pair in a network will select one among several paths for data transfer during a session. Exploring the network and finding the best path (fewer hops, optimum bandwidth, and fewer delays) is the job of routing protocols. The capabilities of available routing protocols depend on the characteristics and requirements of the underlying network for which they were designed.

A routing protocol designed for one type of network may not work properly in other types of networks. In dynamic systems such as ad hoc networks whose characteristics keep changing during operation, finding an optimal path between a source and destination pair is a challenging task for the designer of routing protocols. An optimal solution at time t seconds may become sub-optimal at time $t + n$ seconds; a sub-optimal solution at time t seconds that was dominated by the optimal solution may become optimal at $t + n$ seconds. Continuous exploration and exploration of new and better solutions is essential for routing protocols for dynamic networks.

Different routing protocols employ different search strategies to optimize the solutions they have found. Optimization is the process of designing an algorithm to find the optimal solution for a given problem. In Internet routing, optimization refers to designing the best routing algorithm to achieve maximum efficiency in path selection while optimizing routing parameters such as performance, time, cost of resources and memory consumed by the algorithm. Here the target is to find a resource-rich path in an acceptable time period. As these parameters are scarce in networks, especially in ad hoc networks, the optimization algorithms should use these parameters in an effective way. Optimization problems could be broadly classified as: *deterministic algorithms* and *stochastic algorithms*. In deterministic algorithms, the same steps and state of the algorithm could be repeatable whereas stochastic algorithms will have randomized steps and state [17].

In dynamic networks, such as ad hoc networks, most of the deterministic legacy routing algorithms will not work properly due to their deterministic nature. On the other hand, stochastic algorithms work very well for the dynamic nature of ad hoc networks. Stochastic algorithms may be: heuristic algorithms and meta-heuristic algorithms. A heuristic algorithm works on trial and error basis and gives the best service. Heuristic algorithms hope that the found optimal solution will work most of the time, but not always. An extended version of heuristic algorithms is the meta-heuristic algorithm which normally works better than the heuristic algorithms. Although exact definitions for heuristic and meta-heuristic algorithms could not be found in the literature, the current research trends use the meta-heuristic term for randomized stochastic algorithms. Meta-heuristic algorithms work on two important concepts: *exploitation* and *exploration* and these concepts will exploit the local optimal solution, thereby exploring the global optimal solutions through randomization [17]. The growing popularity of these meta-heuristic algorithms is drawing attention in research because these meta-heuristic algorithms are conversant, effective and easy to understand. Literature has contributed different meta-heuristic algorithms to solve optimization problems. Most of these meta-heuristic algorithms are nature-inspired and mimic the characteristics and the behavior found in insect societies, animal societies and marine mammals.

2.1 Nature-Inspired Routing Algorithms

Nature has found its own solutions for the changing environment, and thereby adapting and surviving unpleasant and dangerous situations. Nature-inspired algorithms mimic nature-found solutions to solve the given optimization problem and often these algorithms use cooperating agents to achieve a global task. These nature-inspired algorithms achieve a diversified solution for a given problem within a reasonable time. Also, with minimal computing effort, these algorithms reach optimal solutions. Another feature of these nature-inspired algorithms is their convergence property. These algorithms use the natural phenomena exhibited by biological systems to adapt themselves to a changing environment as quickly as possible.

Nature-inspired algorithms, also referred to as bio-inspired algorithms, learn and emulate the features of insect swarms, fish schools and other animal groups (herds) for solving complex real world problems. This process is defined as anattempt to design algorithms or distributed problem-solving devices inspired by the collective behavior of social insects and other animal societies [5].

Swarm intelligence (SI), a part of bio-inspired computing, is successful in practice and an influential method used to deal with optimization [10]. SI-based algorithms randomly find multiple solutions for a given problem and then an optimal solution is chosen for the problem. The randomization allows SI based algorithms to search for diversified solutions for a problem. SI-based algorithms exploit good local solution in the current search area and explore global solutions through randomization. SI-based algorithms are simple to understand and easy to implement. Swarms of insects, fish schools and animal herds work in a cooperative manner. Insects use chemicals as messengers for other insects and also to forage. Some animals use the signaling methods for communicating with other animals within the same group. The size and scale of these insect and animal groups will range from a few to thousands. Sometimes these aggregations will have significant influence on both the ecology and human activity.

2.2 Reason for Aggregation

Both intrinsic and extrinsic effects lead to aggregation in insects, fish and other animals. Attraction of other members of the same kind is one intrinsic reason for aggregation (for example, opportunities for mating) and many times the external effects would also cause the aggregation in animal and insect societies (for example, protection from predators, for safe and long migration in birds). Most of the time, for some functional purpose, insects, fish and animals gather together to form a swarm. Fish schools accumulate to get protection from predators and for mating. Bird flocks together to save energy as well as to migrate over longer distance. Animal herds move in a group for coordinated attacks on their prey. But sometimes aggregation has no practical result; aggregation of plankton serves no purpose. In some cases the animals and insects gather together only during their swarming phase and once the purpose is served, they separate. For example, desert locust swarming depends on weather conditions and state of maturity. Behavior of individuals in a swarm is influenced by several factors such as availability of food (quantity and quality), intruder detection and opportunity for mating. [8] [16]

This chapter introduces swarms which are successfully mimicked to solve optimization problems in the real world such as insect swarms (ants and termites), bees, fish school and bird flock by highlighting their inevitable characteristics exhibited while accomplishing a global task. The behavior of other swarms and their analogy with telecommunication networks is little understood in the literature. Hence in this book we concentrate on well proven swarm intelligence techniques. This chapter also discusses the different non-swarm animals that use their cognition to achieve a task. Finally, this chapter gives the analogy between the MANETs routing techniques and the SI techniques to give a clear picture of why SI-based routing algorithms are dominating other classical routing protocols in MANETs.

2.3 Insect Swarms

The foraging of colonies and nest building features of ants and termites exhibit distributed, adaptive and cooperative behavior which could be directly modeled to optimization and control algorithms leading to the ant colony optimization (ACO) and termite optimization terms. In this section the behavioral study of two important insects, namely ants and termites will be discussed.

2.4 Ant Colony Strategy: Finding Shortest Path to Food Source

Through their highly structured behavior, ants can accomplish complex tasks by local means. Finding the most efficient and shortest path to the food source is one of the appealing characteristics of colony of ants which they achieve using stigmergy. Stigmergy is a process of indirect communication which is achieved by chemical substance referred to as pheromones. Ants deposit pheromones during foraging which will be followed and further strengthened by the following ants. Each ant autonomously finds the solution, leaving

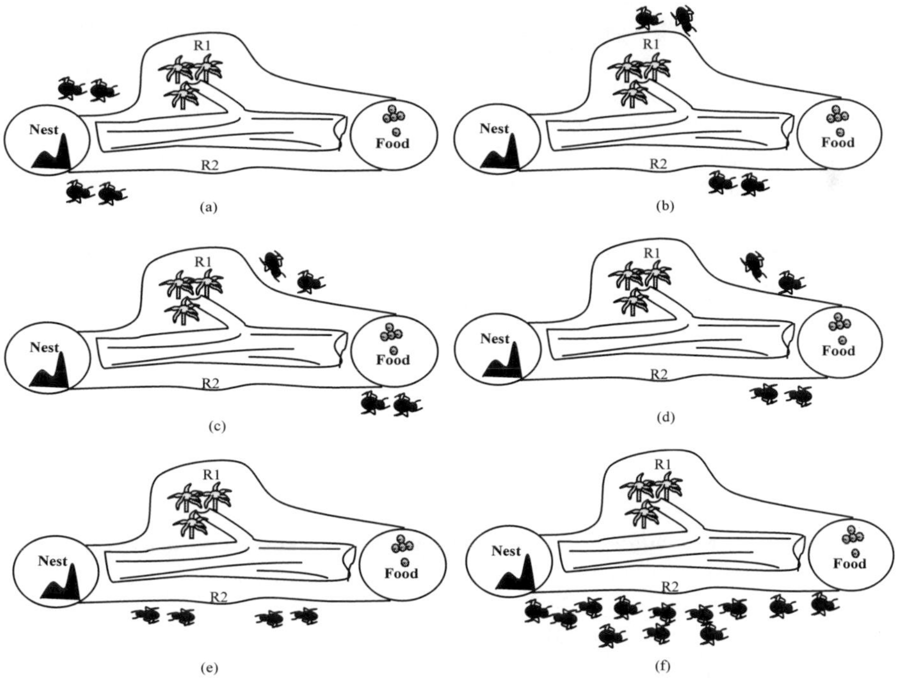

FIGURE 2.1
Ant Colony Strategy: Finding Shortest Path to Food Source

some pheromone for other ants to follow. This independent work of each ant creates many solutions in the beginning with same amount of pheromone applied to each solution. More pheromones are deposited as foraging solutions improve, and better solutions dominates as more pheromones are deposited over time [9].

The problem solving technique of an ant colony in finding the optimal path to the food source from its nest is shown in the Figure 2.1. Suppose there are two routes, *R1* and *R2*, leading to the food source *F*, from the nest *N*, such that *R1* > *R2* in terms of distance to the food source from the nest. In order to make a decision at a choice point if pheromone is present ants will choose the path with higher pheromone concentration; and if pheromone is not present, the arts will makes a random decision among the available paths to the food source. Initially ants will not have any knowledge about the food source and will not find any pheromone trails to the food source; thus they will select the route randomly, either *R1* or *R2*, to search for the food (Figure 2.1a.).

Suppose ants Al and A2 choose *R1* and ants A3 and A4 choose *R2* to search for food. As the ants move along the two routes, they leave certain amounts of pheromones (*pR1* and *pR2*, respectively). Since *R1* >*R2*, ants A3 and A4 will reach the food source before ants Al and A2 on route *R1* (Figures 2.1b and c). To return to the nest from the food source, ants A3 and A4 can choose route $R1$ or $R2$. Because pheromones attract ants and $pR2 > pR1$, ants A3 and A4 are more likely to choose $R2$ over $R1$ (Figure 2.1d).

During their return trip to the nest, ants A3 and A4 will add more pheromones to R2 that will guide future ants to its optimal path. After ants Al and A2 reach the food source and return to the nest, they will find $pR2 > pR1$ and are more likely to select R2 as their route of return to the nest (Figure 2.1e). Future ant groups will choose route R2 to the food source because of its higher concentrations of pheromones (Figure 2.1f).

2.5 Termite Hill Building

Termites are simple and small creatures commonly known as *white ants* since they have ant-like appearance and are autonomous, cooperative and interdependent for their survival. They lack organizational plans and centralized co-ordination. They achieve global tasks by local means through effective co-ordination among them and they can easily recover from setbacks. The hill or mound building nature of termites shows their coordinated behaviour.

The termites use stigmergy, a form of indirect communication, for their co-ordination. The first definition of stigmergy was given by Pierre-Paul Grasse for explaining the behavior of termite societies as *workers stimulated by the performance they have achieved.* The stigmergy is achieved by the termites through a chemical substance called pheromone. The termites construct hills in two phases, a non-coordinated phase followed by coordinated phase. In the first phase, termites deposit pheromone impregnated pellets randomly. Due to non-coordinated behaviour of the termites, many small hills will emerge. Depending on the tenacity and number of the builders, one of the hills will reach the critical size and initiate the second phase. The hill which has reached the second phase will attract more termites through positive feedback and accumulate more and more pheromone. The other hills will remain in the first phase and have lower pheromone gradients. They thus will fail to attract termites. The pheromones on these hills will degrade over time and project negative feedback to termites [9].

Figure 2.2 illustrates the hill building nature of the termite. At $t = i$, due to non-coordinated initial activity, two hills, designated *H1* and *H2*, will emerge quickly (Figure 2.2a).

At $t = j$ (Figure 2.2b), *H1* will reach the second phase and attract more termites. This causes more pellets to be deposited on *H1*, in turn accumulating more and more pheromone. *H2* will be still in the first phase and attracts fewer termites since it has less pheromone concentration. At $t = n$ (Figure 2.2c), all termites will be attracted toward the large pheromone gradient,

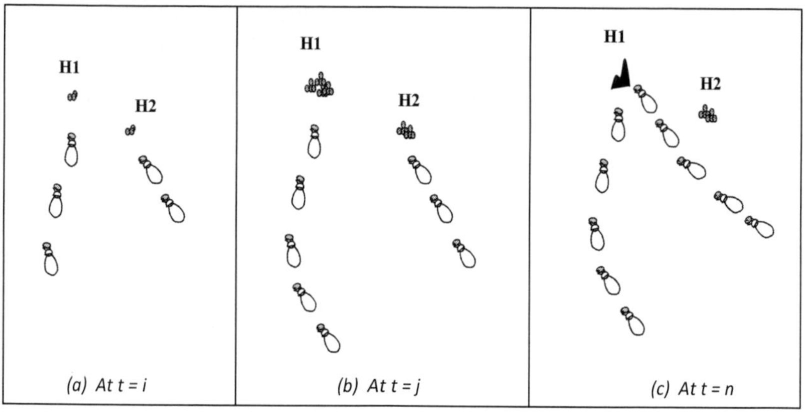

FIGURE 2.2
Hill Building Nature of Termite

that is, *H1*, accumulating more pellets and pheromone. Over time, the hill which has reached the second phase will dominate the other hill and spread a strong positive feedback, eventually forming a sizeable hill. Pheromone over *H2* will be gradually decayed over time, spreading a negative feedback to other termites.

2.6 Honey Bee Colony

Through collective information sharing and processing, bees achieve the most difficult task of finding nectar or pollen supplies and communicating the gathered information to the hive after which workers are recruited to collect the nectar or pollen. Ants, which live in nests over the land, use the pheromone for foraging and spread the pheromone over the land; whereas bees fly in air, cannot use pheromone for communication, and thus use a different kind of communication system. Whenever a worker bee returns to the hive after finding a nectar or pollen source, it will use dancing language to communicate with other workers in the hive. With different dancing styles, a worker bee communicates the distance, direction and quality of the nectar and pollen to other worker bees; it recruits the other worker bees to collect the nectar and pollen. Through dance, these bees convey important information to the other workers in the hive about the distance to and direction of the food. Depending on the distance to the food source, bees perform two round dance or a woggle dance. When the food source is within 50 meters from the hive, a worker bee performs round dance as shown in Figure 2.3a. The worker bee starts from a point x and runs around in a small circle and suddenly it runs in the reverse

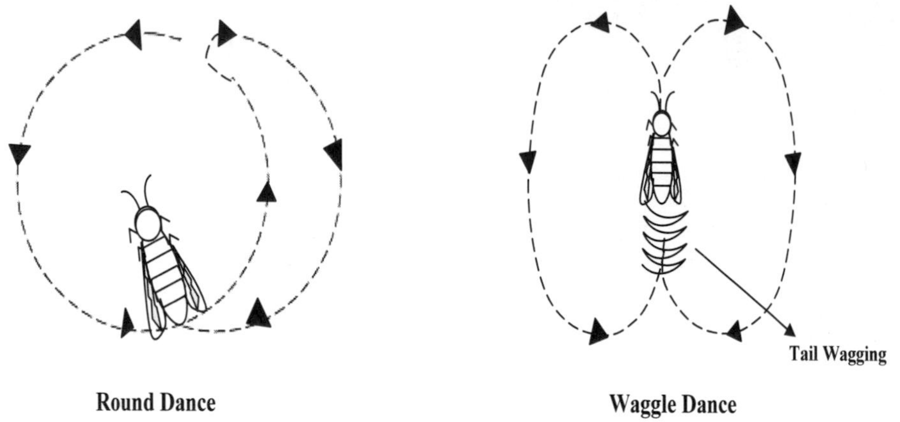

FIGURE 2.3
Honey Bee Dancing Styles

direction and will come back to the source point x. The worker bee will dance several times in different parts of the hive or sometimes will perform in the same place. A round dance informs the other worker bees about the availability of good quality food within the vicinity of the hive. A round dance conveys only the distance but not direction of the food source to the hive.

If a food source is more than 50 meters from the hive, a worker bee will perform another type of dance known as the *waggl* or *wag-tail dance*. The bee starts from a source point designated x and moves in a straight line over a short distance. The bee then returns to the starting point x via a semicircle route, then repeats the steps in an opposite semicircle as shown in Figure 2.3b. During the waggle dance, the worker bee generates a buzz sound and wags its abdomen vigorously from left to right. The other worker bees observe the components of the dance (length of straight line, tempo of the dance, duration of the buzz sound, and duration of dance in seconds). These components are decoded by other worker bees and used to determine the direction and distance from the hive of the food source.

The distance of the food source from the hive is directly proportional to the waggle dance duration (in seconds) by the worker bee. Hence, communicating the distance component to the other worker bees in the hive is very straightforward while indicating the direction of the food source to the hive is convoluted. The worker bee gives the direction of the food source with respect to the position of the sun. The communication of the direction of the food source to the hive can be explained by the following four points:

1. If the food source is in the direction of the sun, the worker bee points its tail wagging dance vertically upward in the hive (Figure 2.4a).

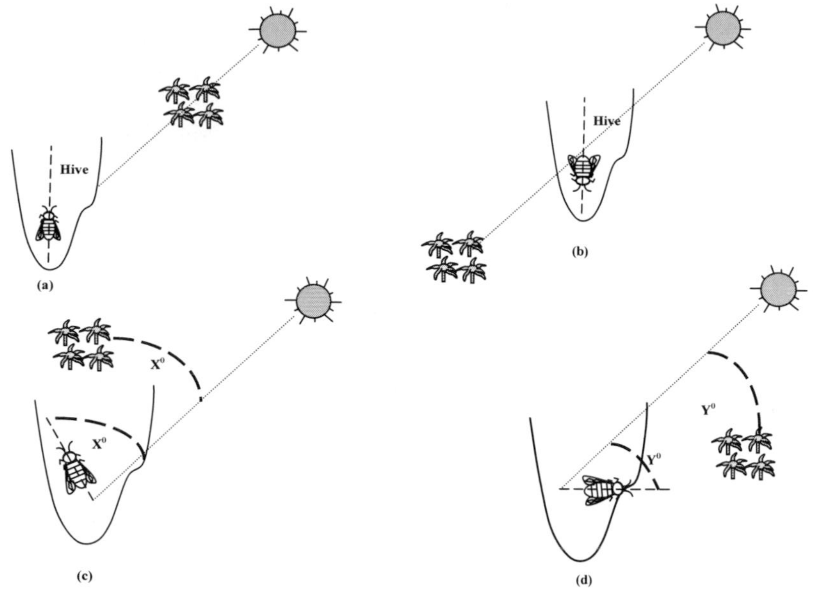

FIGURE 2.4
Different Waggle Dance Styles of Worker Bees

2. If the food source is in the opposite direction of the sun, the worker bee points its tail wagging dance vertically downward in the hive (Figure 2.4b).
3. If the food source is located at an angle x degrees to the left of the sun, the worker bee positions its straight run in wagging dance at an angle x left of the vertical to the hive (Figure 2.4c).
4. If the food source is located at an angle y degrees to the left of the sun, the worker bee positions its straight run in the waggle dance at an angle y right of the vertical to the hive (Figure 2.4d).

Since the sun's direction changes dynamically over time, the direction for the same food source could change according to the sun position and hence the dancing angle by the worker bee must change.

Bee Colony Propagation: Bees show evidence of collective decision making when they select the best site for a new nest among several alternatives. Each scout bee informs the hive about a new nest site by communicating its direction and distance through a waggle dance. Polling and collective decision making determine the new site. As part of the bee biological cycle, colonies rear a new set of queens before they move. One of the new queens remains in the old nest. The old queen and her workers move out of the nest and form a new hive.

Scout bees responsible for finding probable sites for a new nest dance to convey information. Unlike the dance associated with finding nectar or pollen, the hive hunting dance is generally long and energetic. Sometimes a scout bee will stop dancing to monitor the dances of other scouts informing the colony about other sites.

After the components of the dances of all the scout bees are decoded, polling is done and the colony will make a collective decision about the new site. When the swarms travel to the new site, the scout bees lead the group toward the new home. The characteristics of information sharing and processing, polling and collective decision making, and communicating distance and direction via dance language exhibited by bees are fascinating and widely used as optimization techniques for solving real-world problems [15] [6] [4] [11].

2.7 Fish School

A school is a cluster of three or more fish in which each fish persistently fine tunes its speed and direction according to other group members in the school. In a fish school, each fish mimics the movement of all adjoining fish. Fish aggregation is needed to escape from a predator, for breeding or for foraging. In some cases, once the purpose is served, the fish will no longer be members of a swarm while some species of fish will be part of schools throughout their lives. Studies show that at different stages of life, fish aggregate to form schools for example, they aggregate when young and disperse from school when they are adult; they may aggregate for breeding. Fish form schools during morning, and throughout the day they stay in group, and disband at night. The next day they again fish aggregate to form the schools probably with different set of fish. Many a times, different species and different members could be found in the school. Members of the school organize themselves with regular distance and further study reveals that the organization of the fish in the school will save their energy. However, an exact structure of a school is influenced by different components such as speed, size, neighbor visibility, and predator behavior.

School size impacts survival and opportunities for mating. In large groups avoiding predators, collisions and obstructed views create confusion among the fish. Oxygen and food consumption by the leading members of the group leaves trailing groups with inadequate supplies of oxygen and food, leading to unhealthy members. Some fish choose large groups over small groups to foil predators.

The decision to stay in a school or move to another depends on social and practical factors. A fish school may contain several different species. Fish in schools are generally of uniform size because a fish of different size or color may attract predators. Schools usually swim at the same speed. Escape from a predator requires synchronized speed. A member of uneven size or one who cannot accelerate with the group will be attacked by a predator.

2.8 Decision Making by Fish

Fish use different techniques to communicate, for example, visual display, movement of fins, color changes, and sound, to inform other members of a school about defenses, predator presence, migration, mating, and food sources. A fish usually sends its entire life within a specific boundary or may return to a well defined area if it leaves its normal area. Several factors influence a fish to follow other fish but each fish independently makes the decision to follow. Most fish tend to follow the leader of the most attractive fish in a school and thus increase school size [3] [7]. Individuals in schools of fish and flocks f birds follow three rules that government group movement and coordination:

1. Maintain steady distance from the nearest neighbor.
2. Adjust to the same direction as the nearest neighbor.
3. Remain with the group.

2.9 Bird Flocks

Flocking is the synchronized movement of a group of birds. Each bird in the flock follows simple rules to achieve coordinated behavior with its flock mates as shown in Figure 2.5. A bird flock exhibits many characteristics such as coordinated flight among the flock mates, understanding the acceleration of the flock and adjusting to it, and avoiding collision in the flocks. Generally, there will not be any global leader in the flock; each individual will lead the flock at some time [12].

Why do binds flock? Different factors, external as well as internal, cause the birds to aggregate. Moving together reduces the ability of a predator to concentrate on a single individual and attack. Another very important reason for bird flocking is to save energy during long distance travel. The flock head takes the complete responsibility of reducing the up thrust for the other mates in the flock. The flock head flies opposite to down thrust, piercing the air, and

FIGURE 2.5
Bird Flock

moves forward, hence taking the maximum load. This in turn reduces the up thrust required for the other mates in the flock, thereby reducing the amount of energy required to fly. At regular intervals, the exhausted flock head will be replaced with other birds in the flock to keep the momentum of the flock. The new flock head will direct the flock untill its energy is depleted. In this fashion, the bird flock saves its energy during long distance travel. Some of the other reasons for bird aggregation are protection from predators opportunities for mating and finding food, and saving energy during a long flight.

Unlike fish schools in which the size of the school affects some ofthe activities of the school, population does not matter for bird flocks. A flock never becomes oversized. Each individual in a flock concentrates only on itself and its nearest neighbors to ensure that it moves at the same speed and avoids colliding with them. Since all the birds in a flock exhibit the same behavior, the entire flock moves in a synchronized fashion. The only requirement for a new bird to join a flock is for it to move in coordination with the other mates in the flock. All birds in a flock follow three rules:

1. Avoid collision.
2. Stay close to the flock.
3. Adjust the velocity with neighboring flock mates.

Sometimes the flock will split to avoid an obstacle in front of them. Birds in the flock arrange and rearrange themselves to maintain the proper distance with the neighboring flock mates.

2.10 Asymptotic Behavior of Bird Flocks

To study the asymptotic behavior of bird blocks, let us consider Figure 2.6 which illustrates a flock containing n birds designated $Bl, Bl, ...Bn$. We can deduce the following parameters from the figure:

1. Each bird in the flock has a location x_i in the flock.
2. Flock will be moving in the pace p_i.
3. Each bird maintains a distance d with its neighbors in the flock.

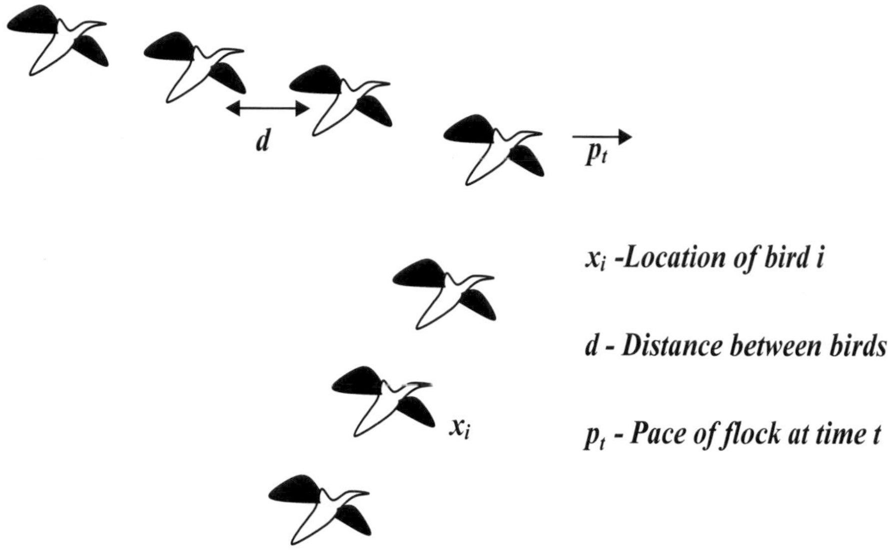

FIGURE 2.6
Bird Flock Behavior

Each bird's location in the flock is distinct; if the location of the bird B_i at time t is x_i, then the location of the same bird B_i at time $t + 1$ can be given as $x_i + P_t'$, where P_t' is a function of neighbor's pace of B_i in the flock. Each bird B adjusts its pace B_p' according to the following equation:

$$B_p' = \frac{\sum_{j=1}^{n} P_i}{n} \tag{2.1}$$

where n is the total number of neighbors of B in the flock. Each bird will fly at a distance d with its neighbors in the flock to avoid the collision in the flock and the birds will maintain the same distance in the flock throughout their journey [2] [12].

2.11 Firefly Algorithm

Most firefly species produce a unique rhythmic flash. As part of their biological behavior, they produce short flashes that they use to communicate with each other, attract prey, or warn others in their group of bitter taste of prey. Rhythmic flashing can be used by males to attract females; females signal interest in mating by flashing the same pattern. In some species, the female mimics the male pattern. The male assumes that the flash pattern means the female is a prospective mate but the female's intent is to catch and eat the male [17].

2.12 Behavioral Study of Non-Swarm Species

In this and the following sections, the behavioral study non-swarm species will be discussed in detail. The cognition used by the mammal bat has inspired many researchers to design optimization algorithms instead of studying non-swarm species such as dolphins and spiders. Also solid theoretical proof for inheriting the principles adapted by these non-swarm species (except bats) is missing in the literature. Hence this section concentrates only on the behavior of mammal bats.

2.13 Bats

The bat is a fascinating mammal known for its advanced and distinguishing echo-location feature; two distinct characteristics of bats make them stand apart from other mammals. They are the only mammals to have wings and the advanced capability of echolocation. The fascinating echolocation feature of bats allows them to find their prey and determine its distance and movement direction in complete darkness.

As shown in Figure 2.7, bats produce a very loud pulse and listen for the echo that bounces back after hitting the prey and surrounding objects. Depending on the species and obstacles, the pulse varies when it echoes back. According to a study, the micro bat species builds a three-dimensional model of its surroundings based on the time delay between the emission of pulses and reception of echoes and also the loudness variations of the echoes. A bat can also detect the type of prey, its distance, orientation, and moving speed using the Doppler effect. Bats also use echolocation to eavesdrop and leak information intentionally or unintentionally [1]. The echoed signals may also be used by other bats and other types of animals to identify nearby objects.

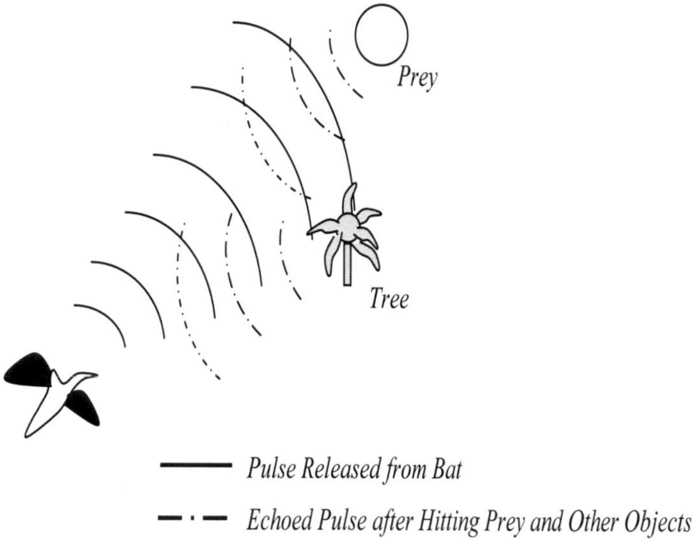

FIGURE 2.7
Echolocation Feature of Bat

2.14 Analogy between SI Techniques and MANETs Routing

Among different swarms in nature, ant and termite characteristics are widely studied and accepted to solve the routing problem in MANETs. Many theoretical models are available to prove the solidity of the algorithms based on ant and termite behaviours in solving routing problems in MANETs. This section, ant and termites correlated features will be with MANETs routing characteristics.

2.15 Ants and MANETs Routing

The main characteristic of the ACO routing algorithm is the acquisition of routing information through path sampling and discovery using virtual ants as small control packets. These ants are generated concurrently by all nodes and operate independently. Their task is to collect information of the quality of a path from the source to the destination or vice versa and update the collected information in routing tables of the intermediate nodes they visit. Reinforcing optimum paths is another task of these virtual ants.

The routing table at node i is derived from pheromone table P^i. It consists of the pheromone scent for each destination d and can reach neighbor node n. Thus $P^i_{n,d}$ refers to the destination pheromone at node i through neighbor node n. These pheromone entries are the local measures of the goodness of the neighbor link to reach destination d and will be continuously updated by reflecting the quality of the path. At each node i, ants stochastically choose next hop nN^i, where N^i represents the total neighbor nodes of i, which will lead to the destination d. While choosing the next hop, the ant gives the preference to higher $P^i_{n,d}$ value which is a function of both pheromone and heuristic values, $P^i_{n,d} = f(P^i_{n,d}, \eta^i_{n,d})$. The η has the same structure as pheromone value and it applies to each neighbor node leading to the destination. For example, number of packets in the queue could be used as a heuristic measure of link goodness. However, not all the ACO algorithms cited in the literature make use of heuristic functions for calculating the selection probability. If it is not used they are only $P^i_{n,d}$ is used for calculating the selection probability.

ACO Characteristics

1. All ACO routing algorithms are adaptive and make use of some incremental learning technique for continuous adaptation.
2. They are multi path routing algorithms.
3. Probabilistic exploratory decisions are integral parts of all ACO routing algorithms.
4. They are simple and emphasize local solutions to achieve global tasks.

To explain ACO algorithms more clearly, the route setup phase of the basic algorithm is discussed in the following sections. We explain how ants make decisions when they face ambiguity and how they update pheromones in the tables in each visited node. Pheromone tables are covered in detail for the reader's convenience.

2.16 Route Discovery Phase of Basic ACO Routing Algorithm

During the route discovery phase, the source node will release n ants to find the optimal path to the destination. While traversing toward the destination node, these ants will deposit the pheromone in each visited node; in this example, it is assumed that each ant will deposit the pheromone value *1* in each visited node. The route discovery phase contains two stages: reverse route setup and forward route setup. The ants which set up the reverse route are called forward ants and the ants which set up the forward route are called backward ants. Ants, while moving from the source node S to the destination node D, will set up the reverse path (path from D to S) and while traversing back from the destination node to the source node they will set up the forward path (from S to D) and this is illustrated in the following example.

Figure 2.8 shows the initial phase of the network where source node s initiates the route discovery phase to find the optimal path to the destination node D. There are two paths from the source node S to destination node D, *R1* and *R2*. As forward ants will not find any destination pheromone trail entry in the pheromone table of S, they randomly choose next hop neighbor nodes to move forward. In this example, two ants will take route *R1* and the remaining two ants will take route *R2*. The pheromone table of node S, which will be empty initially, is also shown in Figure 2.7. It contains three columns, d = *destination*, n = *next hop and* P = *pheromone*. Entries in the pheromone table should be read: *to reach destination node d, next node is n and the pheromone deposited for the next node is P.*

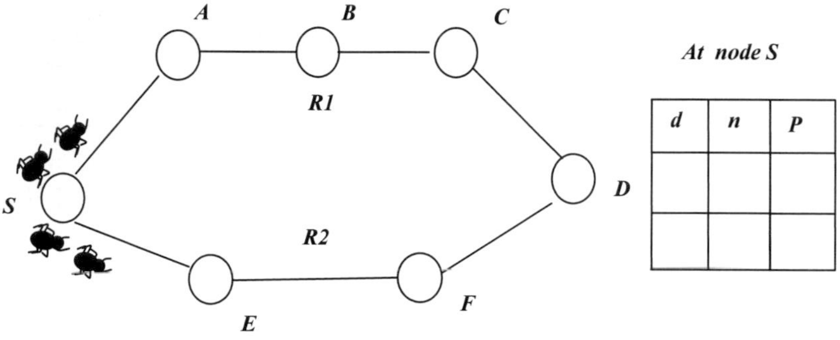

FIGURE 2.8
Example Network

2.17 Reverse Path Setup

During the route discovery, ants reach the next node along their path and also update the pheromone entry in the pheromone table. The pheromone table contents of the node A and E are shown in Figure 2.9(a). At node A, route discovery ants will deposit pheromone for its source S. The first entry in the pheromone table explains key features: to reach source node S, next hop is S and the pheromone deposited for the next node is 2. Figures 2.9(b) and 2.9(c), show the ant position and the pheromone table content of the each node. The solid arrow in the figures tells the path direction which is set up by the forward ants in the pheromone table.

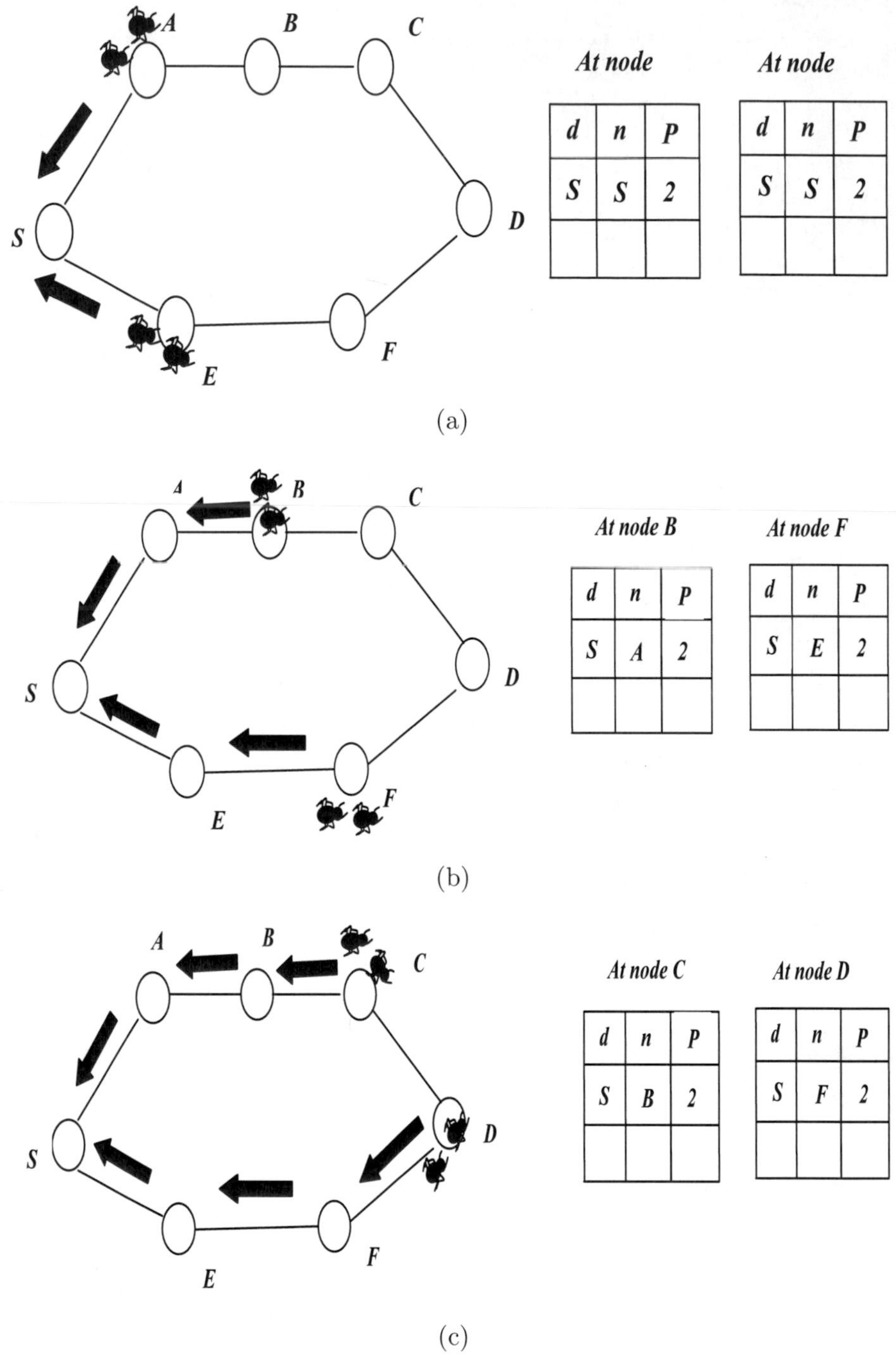

FIGURE 2.9
Reversed Path Setup by ACO Algorithm

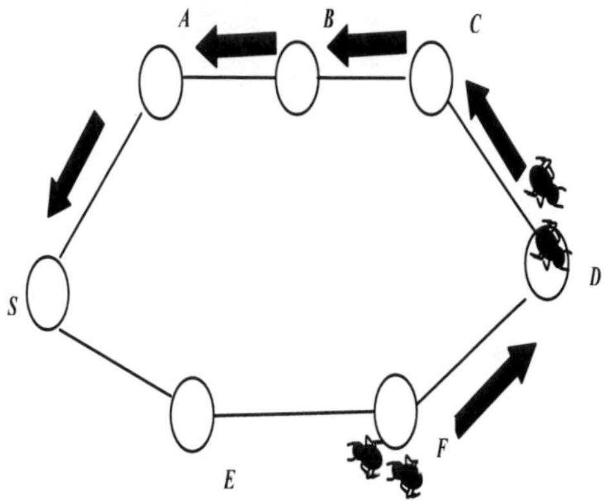

At node D

d	n	P
S	F	2
S	C	2

At node F

d	n	P
S	E	2
D	D	2

(d)

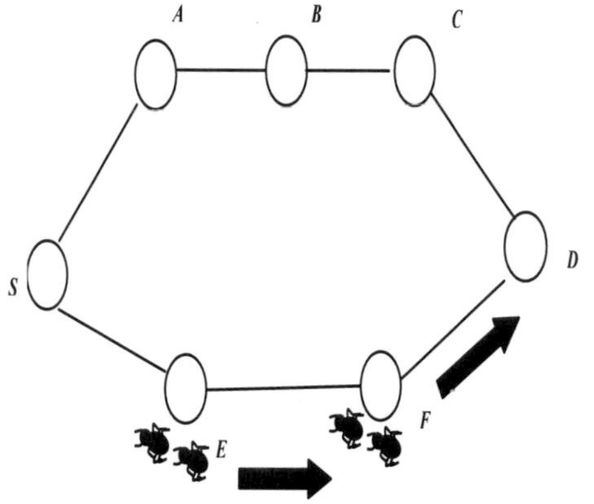

At node F

d	n	P
S	E	2
D	D	4

At node E

d	n	P
S	S	2
D	F	2

(e)

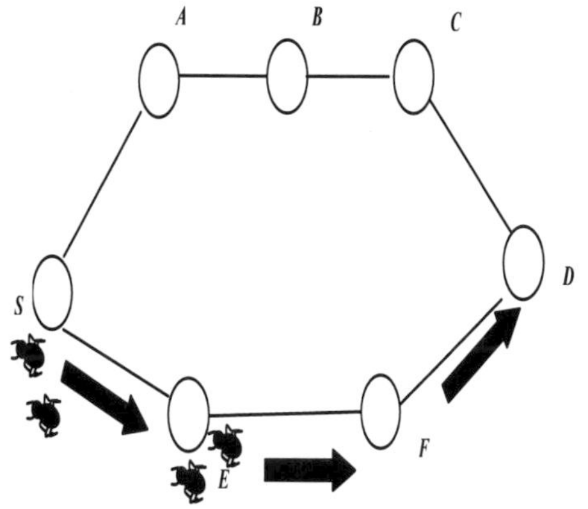

At node S

d	n	P
D	E	2

At node E

d	n	P
S	S	2
D	F	4

(f)

FIGURE 2.9
— Continued

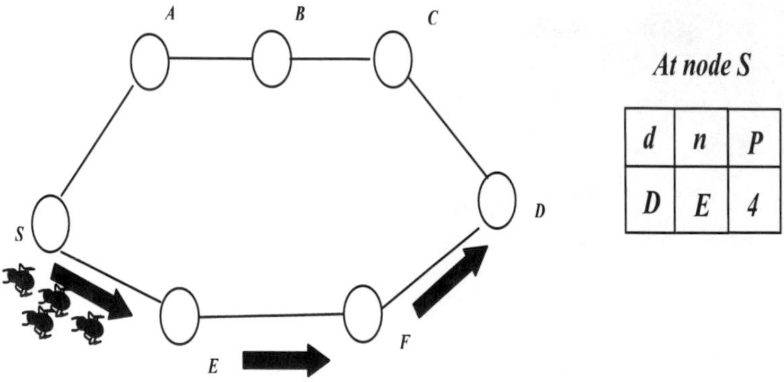

(g)

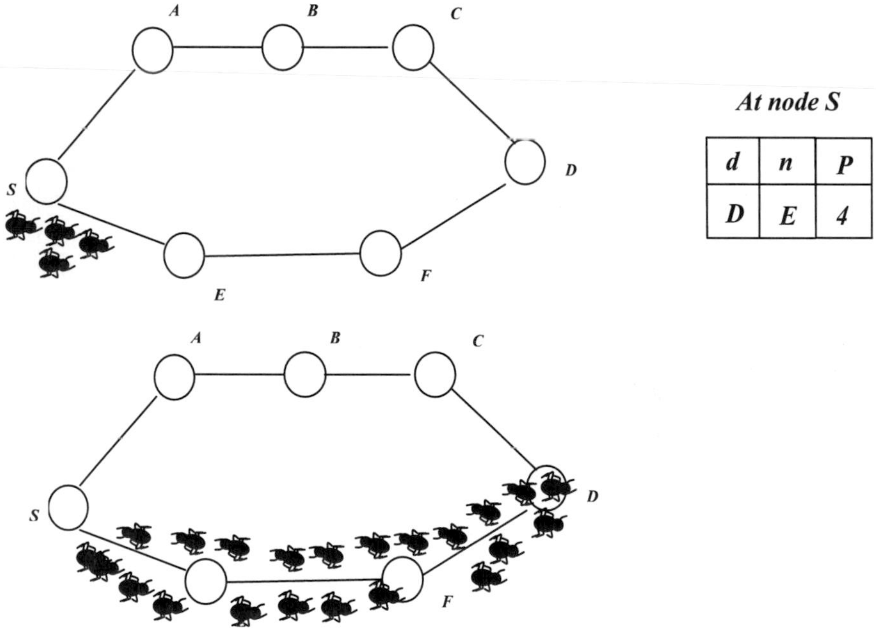

(h)

FIGURE 2.9
— Continued

Once the forward ants over route *R2* reach the destination node *D*, they will traverse back to node *S*. While traversing back, these ants find a source pheromone trail in the pheromone table of node *D*: to reach *S*, the next node should be *F*. Since there is no ambiguity, these ants will follow the pheromone table entry for the *S*, i.e., they will move toward node *F*.

2.18 Forward Path Setup

While returning from the destination, the backward ants will set up the forward path. They deposit the pheromone trail for node *D* in each visited node. In Figure 2.9(d), as the forward ants from route *R1* reach the destination node *D*, they will make another entry for *S* in the *D's* pheromone table. While traversing back from *D* to *S*, these ants find two entries in the pheromone table for *S*, both with the same pheromone value. The ants will randomly choose one among these two alternative paths to reach the source node. Let us assume that ants will choose the first entry in the pheromone table and move toward the node *F* to reach the node *S*. Their positions and the corresponding pheromone table entries in the nodes are shown in Figures 2.9(e), 2.9(f) and 2.9(g).

Once the reverse and forward paths have been se tup, the next foraging ants will find the destination pheromone trail entry in the *S's* pheromone table and accordingly they will follow the path and further strengthen the path by depositing more and more pheromone. This may be observed in Figure 2.9(h). We see that the path chosen by these ants between *S* and *D* is optimal. Hence it shows that utilizing the ant characteristics in finding the optimal path in the networks works very well.

On the other hand, the forward ants on path *R1* have deposited reverse paths in the pheromone tables of nodes *A*, *B* and *C*. As this path will not attract any other foraging ants, pheromone in these tables gradually decays over time and after some time the stale entry will be removed from the pheromone table. Based on the fundamentals of the ACO framework, many optimized ACO-based routing algorithms have been proposed in the literature. These optimized ACO-based routing algorithms are exploratory; they will sample new and good paths continuously and the new explored paths will be kept as backup paths in the pheromone table. The path exploration is kept separate from the data paths so that the data transmission is not disturbed. The up-to-date routing information in the nodes and dynamic adaptation to the context of the network depend on the ant population.

2.19 Termites and MANETs Routing

Inspired by the behavior of social insect termites, an adaptive routing algorithm referred to as the termite algorithm is proposed for MANETs [14] [13]. The packet forwarding technique of MANETs is correlated to the hill building nature of the termite and it achieves strong routing robustness through multiple paths to the destination. As termites work together to build hills, a collection of nodes in MANETs work together to deliver the packet to the desired destination. Like nodes in MANETs, these termites are interdependent, cooperative and there is no central coordinator to control the other termites.

The termite algorithm is an adaptive and per-packet probabilistic routing algorithm where a routing decision is taken for each packet at every node based on the pheromone deposited on the outgoing links. Through stigmergy, it reduces the amount of control traffic, thus increasing the data through put. The termite algorithm treats each mobile node as a termite hill and the packets are strongly attracted toward the hill with highest pheromone gradient. While moving toward the destination, the packets deposit pheromone for their source in each hill visited. This increases the likelihood of packets following the same path while traversing back from the destination. Each node maintains a pheromone table which is analogous to routing tables in traditional routing protocols for tracking the amount of pheromone on each outgoing link. To prevent stale entries in the pheromone table, the concept of pheromone decay over time is introduced. At every node, the pheromone increases linearly and decays exponentially over time. The termite algorithm is driven by three functions, namely pheromone update, pheromone decay and forwarding functions.

In the termite algorithm, the termite workers are bound by the following rules while building a hill:

1. A termite is always attracted toward the pheromone gradient. If no pheromone exists, it moves uniformly and randomly in any direction.
2. Each termite can carry only one pellet at a time.
3. If a termite is not carrying a pellet and encounters a pellet, it will pick it up.
4. If a termite encounters a pellet while carrying one, it will put the pellet down and infuse a certain amount of pheromone on it.

Figure 2.10 show the working principle of a termite algorithm in an example network. The example network consists of two paths, *R1* and *R2*, from source node S to the destination node D. The first phase of hill construction by termite algorithm is shown in Figure 2.10(a). As each node will be treated as a termite hill in the network, the first phase will end up with the small hill in each node on both paths. After j seconds, hills in one of the paths *(R2)* will reach the second phase of hill construction, thereby dominating the hills in the other path *(R1)* and this may be seen in Figure 2.10(b). Finally after n seconds, the path *(R2)* which has reached the second phase will have the bigger hill in each node and spread positive feedback to the other termites as shown in Figure 2.10(c). On the other hand, the path which is still in the first phase will fail to attract the other termites and spreads negative feedback. Pheromone in the hills of the neglected path will gradually decay over time.

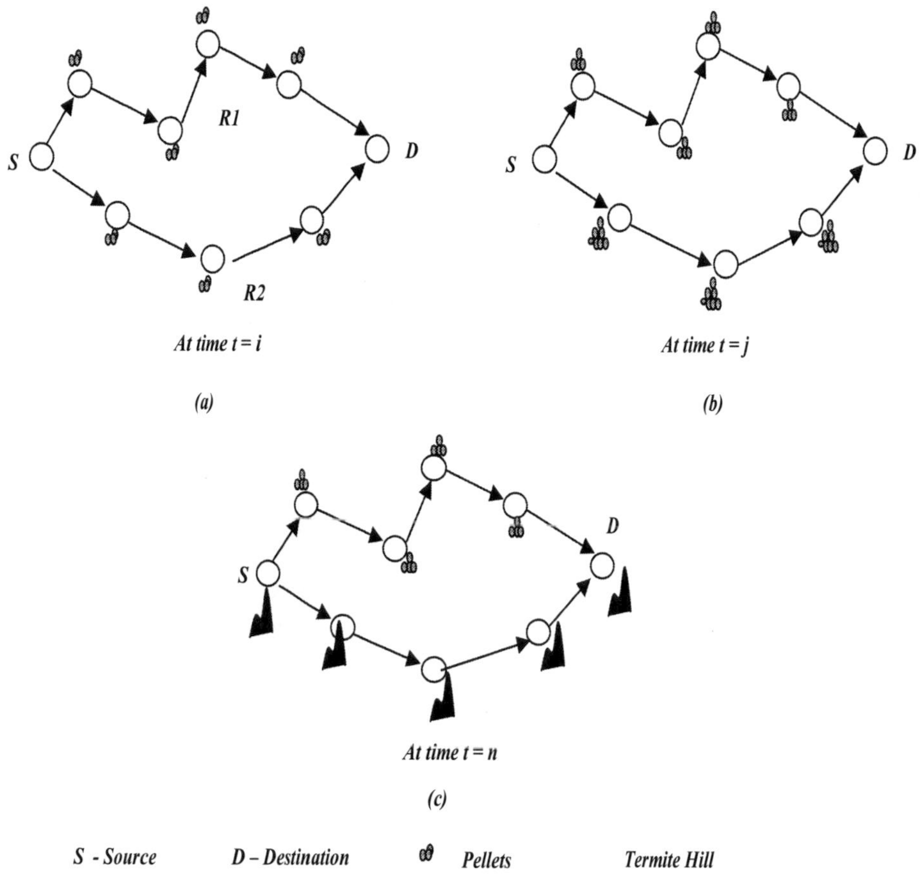

FIGURE 2.10
Using Termite Algorithm to Find Optimal Paths between S and D at Diffrent Time Intervals

2.20 Summary

This chapter starts with a discussion of challenging issues surrounding routing algorithms associated with dynamic networks such as ad hoc types. It highlights the ways in which meta-heuristic algorithms find solutions to the routing challenges of dynamic networks.

We first introduced the inherent characteristics of swarm species that inspired researchers to design and develop novel adaptive routing algorithms for MANETs. The behaviors of swarming species such as ants, termites, bees, bird flocks, fish schools and fireflies and non-swarming species such as bats were discussed. We presented an analogy between ant and termite behaviors and MANETs routing and used an example network to demonstrate how an ant-based algorithm finds the optimal path between source and destination nodes. The progression of a pheromone table in each node is shown for each step. The chapter also details an analogy between termite activity and MANETs routing. We conclude that ant- and termite-based algorithms are capable of finding optimal paths between source and destination nodes in a network.

The next chapter covers SI-based algorithms in detail including their advantages and disadvantages, and the parameters of pheromone update and decay functions for these algorithms.

Glossary

Pellet: Mud particle used by termites to build a hill.

Stigmergy: A form of indirect communication used by social insects such as ants and termites.

Pheromone: A chemical substance used by ants and termites for communication among their own species.

Flock: A group of birds move together in a synchronized fashion.

Exercises

Part A Questions

1. An algorithm that involves randomized steps to achieve optimization is
   - (a) Deterministic
   - (b) Stochastic
   - (c) Deterministic & Stochastic
   - (d) None of the above
2. Meta-heuristic algorithm concept works on
   - (a) Expansion
   - (b) Exploitation
   - (c) Exploration
   - (d) Both (a) and (b)

3. The effects that lead to aggregation in insects, fish and other animals are
   (a) Intrinsic
   (b) Extrinsic
   (c) Intrinsic and Extrinsic
   (d) Animal-centric
4. ACO means
   (a) Animal colony optimization
   (b) Animal common optimization
   (c) Ant colony optimization
   (d) Ant common optimization
5. Stigmergy is achieved by
   (a) Pheromones
   (b) Plankton
   (c) Termites
   (d) None of the above
6. Termites are commonly known as
   (a) Ants
   (b) Fish
   (c) Herds
   (d) White ants
7. The dance performed by honey bees is known as the
   (a) Round dance
   (b) Woggle dance
   (c) Both (a) and (b)
   (d) None of the above
8. If the food is in the opposite direction from the sun, a bee points its tail wagging dance direction
   (a) Vertically upward
   (b) Vertically downward
   (c) Horizontally upward
   (d) Horizontally downward
9. A group of birds moving together in a synchronized fashion is
   (a) Flocking
   (b) Flapping
   (c) Flowing
   (d) Flipping

10. Ants that set up the reverse route in an ACO are called
    (a) Forward ants
    (b) Backward ants
    (c) Both (a) and (b)
    (d) None of the above

Part B Questions

1. Discuss the ant colony strategy in finding the shortest path to the food source.
2. Explain in brief the nature of termite hill building.
3. Explain how a honey bee informs its colony about the distance and direction of the food source.
4. Briefly discuss the inherent characteristics of a fish school.
5. Explain in detail the principle involved in bird flocking.
6. Explain why birds will flock.
7. Explain the route discovery phase of the basic ACO routing algorithm.
8. Explain the analogy between the termite working principle and MANETs routing.
9. For the example network shown in Figure 2.11, give the complete details of route setup by using ACO technique as an example. Consider pheromone update of 0.5 at each node. Give the details of route table content at each node. For route discovery, consider that two ants A and B will choose route $R1$ and ant C and D will choose route $R2$.

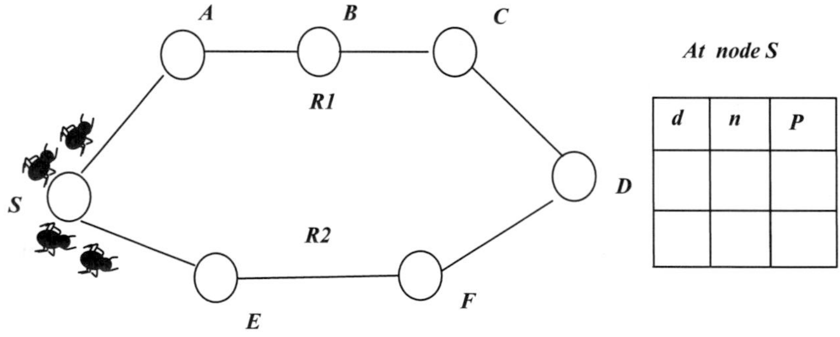

FIGURE 2.11
Question 9 Example Network

10. For the same example network given in network 2.11, give the decay of the pheromone at each node for the time intervals $t = 1, 2, 3...n$. Consider decay of pheromone at each time interval as 0.2.

11. For the example network given in Figure 2.12, show how ants will react when a link fail occurs between the nodes F and D.

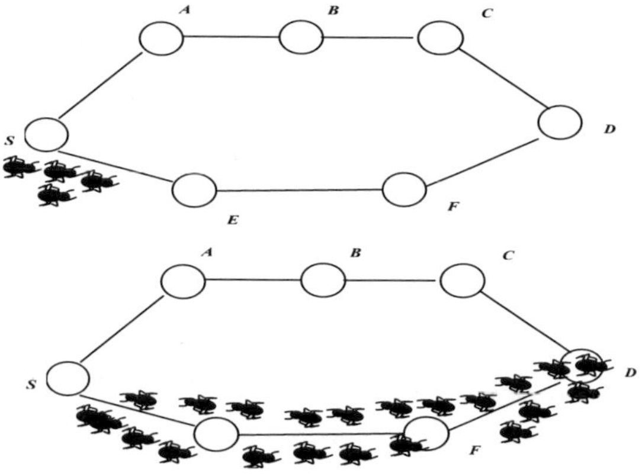

At node S

d	n	P
D	E	4

FIGURE 2.12
Question 11 Example Network

Part A Answers

1. b
2. d
3. c
4. c
5. a
6. d
7. c
8. b
9. a
10. a

References

[1] Matti Airas. Echolocation in bats. In *Proceedings of Spatial Sound Perception and Reproduction*, 2003.

[2] Bernard Chazelle. The convergence of bird flocking. *Journal of the ACM*, 61(4), 2014.

[3] Richard James, Lain D Couzin, David J. T. Sumpter, Jens Krause, and Ashley J.W. Ward. Consensus decision making by fishy. *Current Biology*, 18:1773–1777, 2008.

[4] NC State University David R Tarpy. The honey bee dance language.

[5] M. Dorigo and G. Theraulaz E. Bonabeau. *Swarm Intelligence: From Natural to Artificial Systems.* Oxford University Press, New York, 1999.

[6] KARL VON FRISCH. Decoding the language of the bee. 1973.

[7] Douglas E. Facey, Gene S. Helfman, Bruce B. Collete, and Brian W. Bowen. *The Diversity of Fishes, Biology, Evolution and Ecology, Second Edition.* Wiley-Blackwell, 2009.

[8] Leah edelstein Keshel. Mathematical models of swarming and social aggregation. Thesis University of British Columbia, Vancouver, Canada.

[9] Eric Bonabeau, Marco Dorigo, and Guy Theraulaz. Ant algorithm and stigmergy. *Future Generation Computer Systems*, 16(8):851–71, 2000.

[10] Frank Neumann and Carsten Witt. *Bioinspired Computation in Combinatorial Optimization – Algorithms and Their Computational Complexity.* Natural Computing Series, Springer, 2010.

[11] Eramonn B. Mallon, Nicholas F. Britton, Nigel R. Franks, Stephen C. Pratt, and David J. T. Sumpter. Information flow, opinion polling and collective intelligence in house-hunting social insects. *Philosophical Transactions of Royal Society*, 1567–1583, 2002.

[12] Craig W. Reynolds. Flocks, herds, and schools: a distributed behavioral model. *ACM*, 1987.

[13] Martin Roth and Stephen Wicker. Termite: ad-hoc networking with stigmergy. In *IEEE Global Communication Conference*, volume 5, pages 2937–2941, 2003.

[14] Martin Roth and Stephen Wicker. Termite: emergent ad-hoc networking. In *2nd Mediterranean Workshop on Ad-Hoc Networking.* 2003.

[15] Thomas D Seeley. *The Wisdom of the Hive: The Social Physiology of Honey Bee Colonies.* Harvard University Press, London, 1995.

[16] D. J. T. Sumpter. The principles of collective animal behaviour. *Philosophical Transactions of Royal Society*, pages 5–22, 2006.

[17] Xin-She Yang. *Nature-Inspired Metaheuristic Algorithms, Second Edition.* Luniver Press, 2010.

3

SI Solutions to Routing in MANETs

CONTENTS

In this chapter, existing state-of-the-art bio-inspired routing protocols are discussed in detail while considering several important features of social insects and animal societies. A detailed classification of SI-and non-SI-based routing algorithms is also given based on the code of practice and properties,

Detailed descriptions of bio-inspired routing protocols for MANETs based on the behaviors of ants, bats, bees, birds, and termites are provided. Protocols are examined in relation factors such as congestion, cross layer design, load, location, mobility and quality of service (QoS).

The chapter starts with discussion of legacy routing protocols for MANETs. The well known ad hoc on-demand distance vector (AODV) algorithm and the dynamic source routing (DSR) algorithm are discussed in detail. We explain the drawbacks of legacy routing algorithms in MANETs and reveal how SI concepts correct them. The discussion then focuses on SI- and non-SI-based routing protocols for MANETs and concludes with the taxonomy of SI-based routing algorithms.

3.1 Legacy Routing Protocols

The routing protocols designed for MANETs have to cope with frequent topological changes in order to provide stable paths between the source and destination nodes. The routing protocols designed for MANETs can be categorized as proactive, reactive and hybrid [33] [38]. In proactive or table-driven routing protocols, route updates are propagated throughout the network at regular intervals to maintain the current topology information in each node in the network; thus the main problem with such protocols is maintenance overhead. Hence, proactive routing protocols are not suitable for highly dynamic networks and the examples for proactive routing protocols for MANETs are optimized link state routing (OLSR) [22], destination sequenced distance vector routing (DSDV) [20] and others.

In reactive or source-initiated routing, whenever the source needs a route to the destination, a new route is established between the source and the destination through a route discovery procedure. Route discovery triggers the route request packets to find the optimal path to the unknown destination; thus the control packets are generated only when needed but suffer from high route discovery latency. Dynamic source routing (DSR is one example of this type of routing protocol [23].

Hybrid routing protocols combine the features of both on-demand and table-driven protocols to yield a new class of protocols for MANETs. Efficiently combining the two techniques allows hybrid protocols to improve the performance of a network. One example is the zone routing protocol (ZRP) [19]. The next sections cover ad hoc on-demand routing and DSR protocols.

3.1.1 Ad Hoc On Demand Distance Vector Routing (AODV)

AODV [9], one of the well known reactive routing protocols used in MANETs, is designed and continuously updated by the Internet Engineering Task Force (IETF) for MANETs Working group. Quick adaptation to dynamic topology, low operating cost (processing and memory) and low network utilization are the highlights of AODV. Each node maintains a routing table which consists of only active routes to the destination and gives a quick response to link breakage and topology changes in the network. AODV uses destination sequence number to ensure loop freedom and shows good convergence property. AODV defines three types of messages for route discovery and maintenance, namely route request (RERQ), route reply (RREP) and route error (RERR). Whenever a source node does not find the destination node entry in its routing table to forward a data packet, it initiates the route discovery. In the route discovery procedure, RREQ packets are broadcast to find a fresh path to the intended destination node.

On reception of RREQ packets, a node first updates its routing table to set up the reverse path to the source node and then it either sends RREP packets

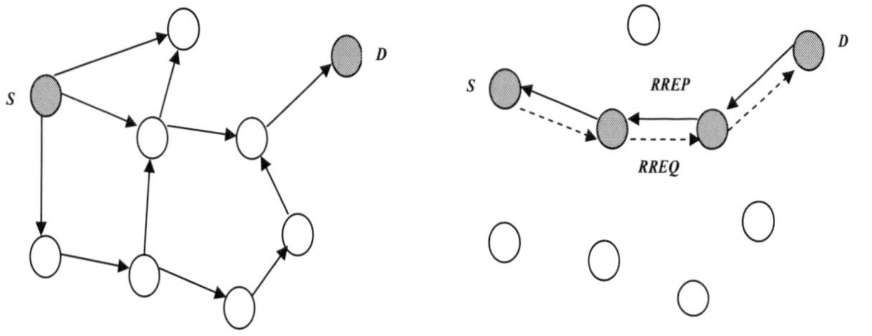

FIGURE 3.1
AODV Working Principle

to the RREQ source or rebroadcasts the received RREQ packet to its neighbors. A node sends RREP packets only under two situations: (1) if the node is the intended destination or (2) if the node finds a destination node entry in its routing table. Otherwise the node will rebroadcast the RREQ packet to its neighbor nodes. RREQ sets up the reverse path to the source on its way while RREP sets up the forward path to the destination. When the source receives the first RREP packet, it initiates the data session by transmitting the stored data packets on the newly found path to the destination. Duplicate RREP and RREQ packets will be discarded by the nodes.

During a data transfer, if a node finds a link break, it sends a RERR packet to the source node to notify the disconnected path; the RERR packet includes the destination address which is not reachable. On the way to the source node, the neighboring nodes will delete the route table entry to the destination node for the corresponding path. Once the source node receives the RERR packet, it reinitiates route discovery to find a fresh route to the destination node. Figure 3.1 shows the route discovery process of AODV routing protocol.

3.1.2 Dynamic Source Routing (DSR)

DSR is another well known routing protocol for MANETs in under the reactive routing protocol category. DSR employs the source routing technique in which the complete sequence of the path is predetermined by the source node. The list of nodes through which the packet is to be forwarded to reach the destination node, is known as the route record and included in the header of each packet by the source node. Upon receiving a packet, a host checks whether it is the indented destination of the packet or not; if the host is the intended destination, the packet is processed accordingly; otherwise, the host simply forwards the packet to the next hop identified in the route record of the packet header.

Whenever a source node does not find the destination node entry in its routing table, it initiates the route discovery process to find a fresh path to the destination node. In the route discovery phase, the source node transmits or floods the RREQ packet to its neighbors to find a new route to the intended destination. Each node, upon receiving the RREQ packet, appends its own identifier in the packet header before forwarding it (Fig 3.1a). When the destination node receives the first RREQ packet, it will unicast the RREP packet to the source node through the route obtained by the RREQ packets header as shown in Fig 3.2b. RREQ packets are flooded while RREP packets are unicast in DSR. The source node, upon receiving the RREP, caches the new route and forwards the buffered data packets. Since the packet header should contain the entire path, the packet size is directly proportional to the path length, i.e., as path length grows, the packet header size also grows. Hence, DSR is suitable for only small networks.

As each node observes the entire route in RREQ and RREP packets, the node also learns new routes to the other destinations and updates its routing table. For example in Figure 3.1a when D receives the RREQ packet from node A, it also learns routes to node B and node C; thus in a single route discovery process many routes will be explored by the intermediate nodes. Further, upon overhearing the RREQ and RREP packets, the neighboring nodes accordingly cache the route record in the routing tables (Figure 3.2.c). Thus this route caching further reduces the route discovery overhead in DSR. The disadvantages of DSR are stale route entries in the routing table which may add extra overhead during the route setup or during the route reconstruction phase. DSR is intended for moderate size network and it works well under low mobility and static network topology; its performance degrades under high mobility. It allows nodes to maintain multiple routes to destinations in their cache, leading to fast route recovery.

3.1.3 Drawbacks of Traditional Routing Algorithms

Traditional routing algorithms do not work proficiently in MANETs because of MANETs distinguishing characteristics. They suffer from high control overhead for discovering and maintaining paths. Path recovery in such algorithms is a complicated process that pauses data transmission for a while and path maintenance is also difficult as the mobility causes frequent changes in the MANETs topology. Hence, designing an efficient, reliable and adaptive routing protocol for MANET is challenging and critical. Thus researchers are working on developing an adaptive routing algorithm as a solution to the above problems of traditional routing algorithms. Designing an efficient routing algorithm for MANETs which copes with its characteristics has been an extensive research effort in recent years.

3.2 Swarm Intelligence-Based Routing

The recent research trends have shown that applying swarm intelligence (SI) principles to find routes in MANETs is giving good results. SI-based routing

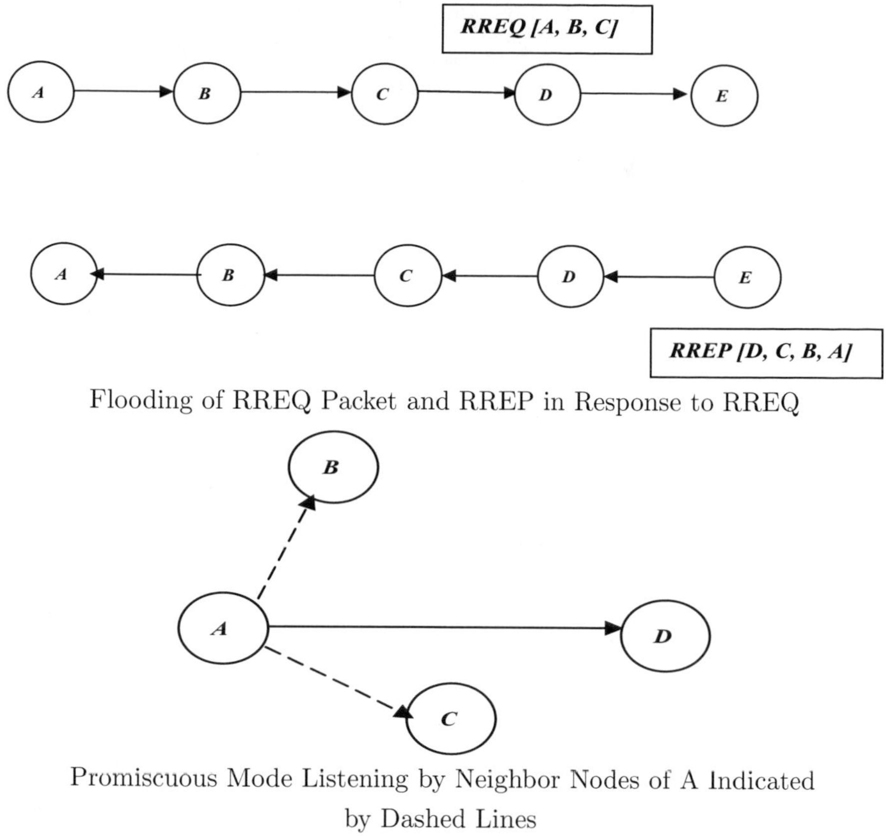

FIGURE 3.2
DSR Working Principle

has been a research topic of great interest. The characteristics and behaviors of swarms such as ants, bees, termites, and schools of fish can be used to design routing protocols for MANETs [16][37] as they show similar characteristics exhibited by MANETs. In SI routing, probabilistic paths are constructed between the source and the destination node using a simple probability update rule. Unlike traditional routing algorithms where explicit paths are set up between the source and the destination node, in SI routing, each route will have a probability of being used. Link failures in such methods could be immediately resolved with the next most probable route, thus reducing control traffic.

Traditional routing protocols always result in same routes in the same whereas SI routing help find the random solution for a system. Traditional routing algorithms are deterministic while SI- based routing algorithms are random and do not have the same properties. Recent research trends have focused on adapting the behaviors of social insects, such as nest building by termites, finding the shortest path to the food source by ants, and energy saving of flocks of birds into a novel routing algorithm for MANETs. SI routing algorithms exhibit adaptive, scalable and robust properties which are also the key routing properties of MANETs routing protocols. Multipath routing, fast route recovery, distributed and Fault tolerance and fast convergence are the advantages of SI-based routing protocols [47].

Ant colony optimization (ACO) is one of the most popular heuristic SI techniques used for solving network optimization problems in general and in particular the routing problem of wireless networks. Accordingly, ACO-based bio-inspired routing algorithms for MANETs are discussed first and then bio-inspired routing protocols are explained in the following sections.

3.2.1 ACO Based Bio-Inspired Routing

The foraging (searching for food) and nest building activities of social insects (ants and termites) indicating cooperative and distributed behavior may be exploited for designing network optimization algorithms [39][51]. The main characteristic of ACOs is frequent acquisition of routing information through path sampling using small independent control packets referred to as ants. The task of these ants is to collect the best path information in both directions from the source to the destination node and accordingly update the routing tables. Schonderwoerd et al. [37] applied the ACO technique referred to as ant based control (ABC) for designing the load-balanced routing protocol for telecommunication networks. AntNet is the first ACO routing algorithm proposed for connectionless networks for the best routing by Di Caro and Dorigo [14]. Inspired by these two algorithms, several ACO-based routing algorithms have been proposed for both wired and wireless communication networks. Di Caro et al. [8] defined an Ant Colony Routing Framework (ACR) for designing ACO-based routing algorithms for wired and wireless communication networks using reinforcement learning and multi agent systems. Existing ACO-based bio-inspired routing algorithms are focused on challenging design issues such as congestion awareness, context awareness, load balancing, and quality of service awareness.

Congestion Aware-ACO Routing Algorithms

Network congestion causes packet drops resulting in reduced throughput; hence congestion aware routing algorithms are required for real-time applications in which satisfactory throughput is maintained at a reasonable level, thereby minimizing packet drops. Several ACO-based congestion aware bio-inspired routing algorithms are proposed for MANETs and the details are given below.

Ad Hoc Networking with Swarm Intelligence (ANSI) (2006)

Rajagopalan and Shen[45] proposed a novel bio-inspired routing algorithm referred to as ad hoc networking with swarm intelligence (ANSI). ANSI uses forward and backward ants for the route setup phase only and then uses data packets for path reinforcement. Forward ants are flooded toward the destination node whenever a new session starts or during the route repair phase. As soon as the first forward ant is received, the destination node will set up the route to the source node and this can be achieved by sending a backward ant. This route setup is evaluated based on the congestion rate of nodes over the single deterministic path through which data packets are forwarded. Further, ANSI uses periodic HELLO packets to refresh the active neighbor list; ANSI also uses the pheromone evaporation concept to negatively reinforce the multiple paths. The authors demonstrated that ANSI performs well when compared to traditional a non-bio-inspired routing protocol (AODV).

Emergent Ad Hoc Routing Algorithm (EARA) (2004)
Liu et al. [55] proposed a novel bio-inspired multipath routing concept known as the emergent ad hoc routing algorithm (EARA). The algorithm uses ants like agents to find new paths locally. During route discovery, these agents take random walks across the network to find multiple routes to a destination. The optimal and sub-optimal paths are marked through pheromone trails. EARA uses local foraging ants (LFAs) to find new routes when all the pheromone trails of the destination drop below threshold level. Periodic refreshment of the neighbor list can be carried out by HELLO packages. The agents continuously explore new and better paths during data sessions to find the best available path for data transfer. The algorithm uses the pheromone and a heuristic value (congestion in a neighbor node) to calculate the probability of the next hop. The results demonstrate that EARA met the requirements of a dynamic network without additional control packet overhead very well.

AntHocNet (2005)
Ducatelle et al. [15](2005) proposed a novel bio-inspired routing algorithm for MANETs. AntHocNet uses reactive forward and backward ants to discover new routes. These ants are sent over high priority queues and at each node these ants are either unicast or broadcast to collect the network information. Whenever a forward ant reaches the destination, it will be converted into a backward ant and follow the same route in the reverse order by updating the pheromone table in each visited node. The pheromone update is proportional to the number of hops in the path, traffic congestion and signal to noise ratio. AntHocNet continuously finds new paths to the destination using proactive path maintenance and path exploration mechanisms, thereby removing old (unused) paths using slow pheromone diffusion. AntHocNet performance has been extensively evaluated through simulation under different scenarios and it has been shown to outperform the classical AODV and OLSR routing protocols. Kalavathi and Duraiswamy [3] have shown little improvement to AntHocNet by adding a node disjoint multipath feature.

Ant Routing Algorithm Based on Adaptive Improvement (ARAAI) (2005)
Zeng and Xiang [54] proposed a bio-inspired multipath routing algorithm designated ARAAI. It uses forward and backward ants for the route discovery

and setup phases. The forward ants collect local node and path information, thereby finding routes to a destination. For each forward ant reaching the destination, one backward ant is released and establishes multiple paths to the destination. The next hop is chosen based on the pheromone and local heuristic (link stability) value of the link. Periodic HELLO packets are used to refresh neighbor tables. ARAAI targets stagnation (network reaches single link dominant state) by using adaptive parameter coordination. Results have shown that ARAAI is more efficient than the AODV and DSR protocols.

Probabilistic Ant Routing (2009)

Prasad et al. [36] proposed a bio-inspired routing algorithm referred to as PAR which uses forward ants (FANTs) for exploring the routes and network information whereas backward ants (BANTs) are used for updating the data structures in the nodes. A FANT uses the same queue as the data packet but a BANT needs high priority queues. On reaching the destination, the FANT transfers the collected information to the BANT and the destination node will remove the FANT from the network. The BANT traces the path of the FANT to update the data structures and the routing table in each visited node. Ants will be either unicast or broadcast, depending on the pheromone information over the neighbor links and periodic single-hop HELLO packets are used by every node to maintain the fresh neighbor list. Simulation results have demonstrated that PAR performs better than AODV.

Cross Layer Design-Based ACO Routing Algorithms

Since the traditional protocol stacks will not function efficiently for MANETs, several efforts have been carried out by many researchers to improve the performance of protocol stacks using the cross layer design (CLD) and thereby providing stack-wise network interdependence. CLD has attracted several researchers for developing adaptive quality of service (QoS) aware routing protocols for MANETs and the details of CLD-based ACO routing algorithms are given below.

EARA-QoS (2005)

Liu et al. [56] developed a CLD based QoS aware ACO routing algorithm referred to as EARA-QoS. It was designed to improve performance metrics such as delay and congestion and provide the best quality service using CLD across MAC and network layers. EARA-QoS uses two heuristics values (delay and congestion) for finding the next-hop probability to forward the data packets; further it uses sequence numbers to avoid the loops and provide multiple paths between the source and destination. Simulation results have demonstrated that EARA-QoS outperforms the traditional AODV routing protocol.

Load Aware ACO Routing Algorithms

The existing routing algorithms for MANETs use routing metrics such as congestion, node mobility, and end-to-end delay in general and in particular the route hop count is the most widely used metric. The nodes which appear in the shortest path will get heavy load and thus exhaust the available network resources such as bandwidth, energy, memory and this creates a bottleneck in the shortest path resulting in-congestion problem. Since mobile nodes in MANETs have limited capacity, thus there is a need for designing load-balanced routing algorithms. State-of-the-art load aware ACO-based bio-inspired routing algorithms for MANETs are explained below.

Multi-Agent Ant Based Routing Algorithm (MARA) (2007)
Sivakumar and Bhuvaneshwaran S [12] proposed a multi-agent load aware ACO routing algorithm referred to as MARA uses route discovery, route update, data routing, route maintenance and route failure handling phases for achieving the load-balance. MARA uses forward and backward ants for route discovery and setup phases and each forward ant shows visited nodes. Whenever the destination node receives the forward ant, it discards the duplicate forward ants and generates the backward ants for updating the pheromone quality along the path. More paths are explored and added in the routing table using a proactive path maintenance scheme; thus MARA achieves load balancing through a probabilistic routing strategy.

AntOR (2010)
Villalba et al. [24] developed a load aware ACO routing algorithm referred to as AntOR which is an extended version of AntHocNet with three additional features namely: (1) disjoint link and disjoint node routes between the source and destination; (2) best paths exploration based on the distance in terms of number of hops; and (3) two types of pheromones, namely regular pheromone used for denoting the path for data sessions and the virtual pheromone used for denoting the possibly good path for data sessions. Through simulation results the authors demonstrated that AntOR outperforms AntHocNet due to its load balancing feature.

Mobility Aware ACO Routing Algorithms
The mobility of nodes makes MANETs topology dynamic in nature and results in difficult route setup and maintenance phases. Further, MANETs also suffer from frequent link failures due to the mobility of nodes, thereby causing a source node to spend most of its time in route setup and maintenance, resulting in low throughput and more control packet overhead. Thus there is a need for mobility aware routing algorithms for MANETs and the details of a mobility aware ACO-based bio-inspired routing algorithm are as follows.

Ant Colony Optimization for Mobile Ad Hoc Networks (PACONET) (2008)
Eseosa et al. [35] proposed a mobility aware ACO-based bio-inspired routing algorithm referred to as PACONET which uses forward and backward ants for route discovery and setup. Forward ants are broadcast in a controlled manner to discover new routes and backward ants are used for establishing the path. Forward ants take the path of unvisited nodes thus make sure that all possible paths to the destination node are discovered; data packets are routed

stochastically on the highest pheromone link. Through simulation, PACONET has proven to be better than non bio-inspired AODV.

Location Aware ACO Routing Algorithms

The dynamic topology of MANETs causes frequent link breakups that cause the source node to spend most of its time in route setup and maintenance. In position-based routing algorithms, each node will have a general idea about the network topology and its neighbors and can choose the nearest neighbor toward the destination. Hence, dynamic networks like MANETs require location aware routing protocols and the details of the state-of-the-art location aware ACO-based bio-inspired routing algorithms are described below.

GPS and Ant-Like Algorithm (GPSAL) (2001)

Camara and Loureiro [6][7] developed a Global Positioning System (GPS)-based ACO routing algorithm referred to as GPSAL which works on the assumption of the presence of GPS-based mobile devices. The forward ants are flooded toward remote nodes to collect and disseminate the routing information and ensure bandwidth efficiency. GPSAL was compared with non-bio-inspired location aided routing (LAR) and their results demonstrated that GPSAL requires less routing overhead when compared to LAR.

Mobile Ant-Based Routing (MABR) (2006)

Heissenbuttel et al. [21] designed and developed a novel location aware ACO-based bio-inspired routing algorithm for large scale ad hoc networks with irregular topologies derived from the ant-based mobile routing architecture (AMRA). MABR divides the network area into rectangular zones and uses the concept of a logical router for long distance routing. Ants are used to update the routing tables proactively in these logical routers and position-based routing protocols are used for data forwarding. Further, pheromone evaporation of biological ants is mimicked to remove the stale path entries. The authors proved that MABR achieves superior performance when compared to GFG/GPSR.

Ant Routing with Distributed Geographical Localization (2009)

Kudelski and Pacut [25] proposed an extended version of AntHocNet with distributed geographical localization knowledge for MANETs. The proposed algorithm partitions the network area into cells and routing is considered at the cell level to provide the knowledge locally. The authors demonstrated that the proposed algorithm outperforms AnthocNet.

HOPNET (2009)

Jianping et al. [50] proposed a location aware ACO-based routing algorithm referred to as HOPNET which combines the ACO principles with zone routing protocol (ZRP) and DSR protocols; HOPNET is based on hopping of ants from one zone to the next zone. The network is divided into routing zones according to the radius measured in hops which may vary in size; route discovery is locality proactive within the zone and reactive across the zones.

HOPNET maintains two routing tables in each node, namely IntraRT (intra zone routing table which will be maintained proactively) and InterRT (inter zone routing table which will be maintained reactively). In IntraRT, HOPNET uses ACO techniques and periodically sends forward ants to explore the paths within the zone and backward ants are sent with the discovered path. Ants in IntraRT are classified into five types: internal forward, external forward, backward, notification, and error. The InterRT uses the DSR technique to store routes across zones. Simulation results show that HOPNET is highly scalable for large networks when compared to the AntHocNet routing protocol.

Hybrid Ant Colony- and ZHLS-Based Routing Protocol (HRAZHLS) (2010)
Rafsanjani et al. [28] devised a location-based ant routing algorithm called HRAZHLS that combines the features of both ant colony and zone based hierarchical link state (ZHLS) routing protocols. The network area is divided into non-overlapping zones where each zone size is dependent on node mobility, network density, transmission power and propagation characteristics. Like HOPNET, HRAZHLS also maintains two routing tables in each node namely, IntraRT and InterRT. It also uses the same five types of ants used in HOPNET. It uses proactive approach for interzone route maintenance whereas the reactive approach is used for intrazone route maintenance. The authors claim that the proposed algorithm improves the packet delivery ratio, and reduces the delay and overhead.

Automatic Clustering Inspired by Ant Dynamics (ACAD) (2012)
Aritra and Swagatam [2] proposed an automatic clustering-based heuristic algorithm referred to as ACAD which detects the separated clusters of any shape, either convex or non-convex, using pseudo ants. Pheromone decay is also used to remove the old solutions and the authors demonstrated that ACAD exhibits promising results with real and synthetic data sets.

Other ACO Routing Protocols for MANETs
In this section other ACO based routing algorithms are highlighted by focusing on performance optimization in terms of throughput and end-to-end delay and do not fall under any category discussed above. These algorithms use ACO techniques mainly for convergence or for route dissemination.

Cooperating Mobile Agents for Dynamic Network Routing (1999)
Minar et al. [31] developed an ACO-based routing framework referred to as cooperating mobile agents for dynamic network routing which generates the mobile agents independent of network and data session at the network setup. These agents do not die and continuously take random walks across the network to maintain the history of previous visited nodes and updates the same whenever a new node is visited. Through simulation results authors have demonstrated that the proposed framework exhibits superior performance under highly mobile network conditions.

Accelerated Ant Routing (2001)

Matsuo and Mori [18] developed a ACO-based *accelerated ants routing* algorithm for MANETs referred to as AAR and derived from ARA [57]. AAR concentrates on convergence property and no return rule (previous hop is not chosen as the next hop to avoid the network loop); it uses N step backward exploration to explore the network effectively. For path exploration, AAR uses regular ants which consist of a stack of last N visited nodes and accordingly updates the pheromone table of each visited node. AAR uses both pheromone and heuristic values (queue values) to decide the next hop to forward the data packet to the destination node. The authors have shown that AAR achieves superior performance and convergence when compared to AntNet, Q-routing and PQ-routing.

Ant Colony-Based Routing Algorithm (ARA) (2002)

Gunes et al. [26] developed an ACO-based routing algorithm referred to as ARA which inherits the features of both AODV and AntNet and uses both forward and backward ants for route discovery and setup. ARA uses data packets to update the pheromone table and thus reduces the ant population for continuous path sampling. The pheromone evaporation concept of real ants is also mimicked to remove the old data entries in the table. Simulation results have shown that ARA performs better than the traditional routing protocols such as AODV, DSDV and DSR under low and moderate dynamic topological conditions.

Ant-AODV (2002)

Marwaha et al. [29] proposed an ACO-based routing algorithm for MANETs referred to as Ant-AODV in which a proactive route update mechanism is exploited along with the salient features of the traditional reactive AODV routing protocol. Ant-AODV consists of ants which randomly traverse the network and keep a record of last N visited nodes and accordingly these records are updated in each visited node. Since ants are used to continuously monitor and explore the new routes, Ant-AODV increases the network connectivity by reducing the route discovery latency. Ants continuously explore new and better routes; data packets are routed to the best available path to the destination and thereby decrease the end-to-end delay. Data packets are routed as per the AODV norms and ants are used only for spreading the routing information. The authors have shown that Ant-AODV achieves the better network connectivity with low end-to-end delay when compared to the AODV routing protocol.

Probabilistic Emergent Routing Algorithm (PERA) (2003)

Baras and Mehta [5] developed an ACO-based routing algorithm referred to as PERA which floods the forward ants only during the beginning of the communication session or when routes are stale for route discovery. For each forward ant reaching the destination, backward ants are launched, thus leading to multiple paths between the source and destination. Backward ants update *routing* and *statistic* tables in each visited node. A reinforcement parameter *(either delay or number of hops)* is also used during the table updating to reflect the current state of the network. Data transmission takes place in the best path available and the rest of the paths are considered for route recovery. Simulation results have shown that PERA is better than AODV.

ABC-AdHoc (2004)
Tatomir and Rothkrantz [4] proposed a hybrid routing algorithm based on ant based control and AntNet and called ABC-AdHoc. ABC-AdHoc uses the forward ant concept for updating the pheromone over the path and uses AntNets framework for probabilistic routing decisions. Through simulation results the authors demonstrated that ABC-AdHoc achieves superior performance when compared to AntNet.

W_AntNet (2007)
Dhillon et al. [13] proposed an AntNet-based routing algorithm for MANETs referred to as W_AntNet and it uses beacon messages for periodic neighbor discovery. W_AntNet is dependent on buffer size of node and exhibits satisfactory performance under static network conditions. W_AntNet suffers from looping packets and deteriorated performance under the dynamic network conditions when compared to AODV and DSR routing protocols.

Enhanced Ant Colony Based Algorithm (2008)
Cauvery and Vishwanatha [34] proposed an enhanced ACO-based routing algorithm for MAENTs which is an extended version of the ARA algorithm [28] for generating all possible paths between the source and destination nodes. Periodic route refreshing is used to find new and better paths; a time-out mechanism is also used to resend the ants to deal with ant loss during the route discovery. Memory buffer is used to hold the packets during link failure, thereby avoiding packet retransmission. The proposed algorithm was evaluated with static data and its performance was better than ARA.

Scented Node Protocol (SNP)(2010)
Sagduyu et al. [52] proposed an ACO-based routing protocol SNP for MANETs which uses odor localization and tracking of insect colonies to enhance the routing. SNP gives scent-based guidance to ants for route discovery and maintenance. Whenever a source wants to find a route to the destination, it sends the broadcast ant towards, the destination; each destination will have a food scent that attracts these ants. If an ant finds the destination scent, it will change its communication from broadcast to unicast mode. HELLO messages are used for spreading the scent information and thus HELLO messages help in local route repair. The authors have shown that SNP exhibits superior throughput when compared to AODV and AntHocNet.

Simple Ant Routing Algorithm (SARA) (2010)
Correia, and Vazao, [10][11], proposed a simple ant routing algorithm for MANETs referred to as SARA which optimizes the routing process by

using controlled neighbors broadcast (CNB), control message (FANT) and deep search strategies. The simulation results have shown that SARA achieves the lowest overhead while increasing the throughput.

3.2.2 Bees-Based Routing Algorithms

A honey bee colony can intelligently choose the quality and quantity of the food source through simple behavior of individual bees. A honey bee colony can contain several kinds of worker bees such as food storer, scout and forager which are responsible for recruiting and foraging-Bees perform dance (waggle dance and tremble dance) to communicate with their fellow bees-State-of-the-art bee based bio-inspired routing algorithms for MANETs are discussed below.

BeeAdHoc Routing Protocol (2005)
Wedde and Farooq [46] developed an energy efficient bee-inspired routing algorithm for MANETs referred to as BeeAdHoc by using the foraging principles of honey bees. BeeAdHoc uses scouts and foragers for route establishment and maintenance phases of MANETs. Through extensive simulation, the authors demonstrated that BeeAdHoc exhibits superior performance in terms of packet delivery ratio, delay and throughput with lower energy consumption.

BeeAIS Routing Protocol (2007)
Mazhar and Farooq [30] have developed an extended version of BeeAdHoc for MANETs. BeeAIS uses the artificial immune systems (AIS) framework to detect the misbehavior in BeeAdHoc and thereby provides security. The authors have shown that BeeAIS can counter the routing attacks in BeeAdHoc efficiently.

3.2.3 Flocks of Birds-Based Routing

Birds travel in flocks of V shape where the sphere head takes the entire burden and reduces the up thrust required by the rest of the birds in the flock; thus the other birds in the flocks save energy. When the sphere head loses its energy, it is replaced by another bird to take the burden of the flock. Each bird in the flock will not have any idea about its position in the flock and no specific bird directs the movement. Thus birds can travel a long distance without losing much of energy and this feature inspired many researchers to develop energy efficient bio-inspired routing algorithms for MANETs and the details are given below.

Bird Flight Inspired Clustering-Based Routing Protocol (2010)
Tiwari and Varma [32] proposed a bird flight-inspired clustering-based scalable and energy efficient routing protocol for MANETs. The proposed algorithm actively maintains the network clusters and uses position-based routing to

choose dynamic gateway nodes and thus reduce the flooding of control packets. Through simulation results, the authors have shown that the proposed algorithm exhibits superior performance.

Bird Flocking Behavior Based Routing (BFBR) (2006)
Srinivas et al. [48] developed a bird flocking behavior-based routing for MANETs referred to as BFBR which uses encounter search to improve the route discovery latency; BFBR uses direction forward routing for effective route maintenance. Through simulation results the authors have demonstrated that BFBR exhibits encouraging results.

3.2.4 Termite-Based Routing Algorithms

Termites, also known as white ants are autonomous, cooperative and interdependent social insects that lack organization plans and centralized coordination. They achieve global tasks by local means through effective mutual coordination and can easily recover from setbacks. Termites use the stigmergy concept via a chemical substance known as a pheromone. The hill building nature of termites demonstrates their coordinated behavior. They construct hills in non-coordinated and coordinated phases. In the first phase, they randomly deposit pheromone-impregnated pellets. Many hills will emerge initially because the termite activities are uncoordinated (Figure 3.Sa). Based on density and critical size, one of the hills will reach the second phase (Figure 3.Sb). The larger hill will attract more termites and in turn accumulate a large pheromone gradient that will spread strong positive feedback. Figure 3.5c depicts the decay of the pheromones deposited on the other hills over time, causing them to emit negative feedback [27]. Inspired by the behaviors of social termites, an adaptive routing algorithm known as the termite algorithm was proposed for MANETs [40] [41].

The packet forwarding technique of MANETs is correlated to the hill building activities of termites and achieves routing robustness through multiple paths to the destination. The termite algorithm is adaptive and features per-packet probabilistic routing. A routing decision is made for each packet at every node based on the amount of pheromone deposited on the outgoing links.

Through stigmergy, the termite algorithm reduces the amount of control traffic, thus increasing data throughput. The algorithm treats each mobile node as a termite hill and the packets are strongly attracted toward the hill with the highest pheromone gradient. As the packets move toward the destination, they deposit pheromone for their source on each hill they visit. This increases the likelihood that packets will follow the same path as they return from the destination.

Each node maintains a pheromone table (analogous to routing tables in a traditional protocol) for tracking the amount of pheromone on each outgoing link. The concept of pheromone decay over time noted in termite hills

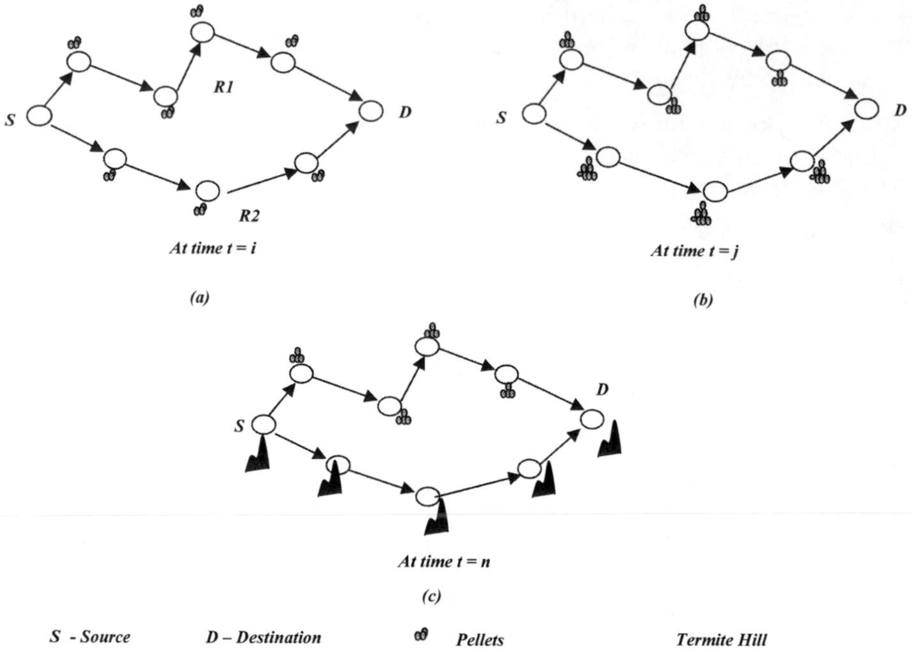

FIGURE 3.3
Hill Building Nature of Termites at Different Time Intervals

is introduced to eliminate stale entries in the table. The pheromone increase at every node is linear and decays exponentially over time. The termite algorithm is driven by three functions: pheromone update, pheromone decay, and forwarding.

Pheromone Update: Whenever a node receives a packet, it increments the pheromone for the packet source by a constant γ in the pheromone table. The pheromone update equation of the termite algorithm is shown in Equation (3.1) where s is the source node; i is the previous hop and $P'_{i,s}$ is the updated pheromone value in the pheromone table:

$$P'_{i,s} = P_{i,s} + \gamma \tag{3.1}$$

Pheromone Decay: To remove old solutions periodically (at every 1 sec), each entry in the pheromone table is multiplied by the decay factor $e^{-\tau}$, where τ is the decay rate. The pheromone decay equation is:(3.2)

$$P'_{n,d} = P_{n,d} \cdot e^{-\tau} \tag{3.2}$$

Forwarding: Based on the amount of destination pheromone deposited over the neighbor link, the probability of neighbor link usage is calculated using forwarding function:(3.3).

$$p_{n,d} = \frac{(P_{n,d} + K)^F)}{\sum_{i=1}^{N}(P_{n,d} + K)^F} \tag{3.3}$$

where $p_{n,d}$ is the probability of using neighbor node n to reach destination d and N represents the total neighbor nodes for node n. Constants K and F denote the pheromone threshold and pheromone sensitivity which are used to tune the routing behaviour of the Termite algorithm. Simulation results of the algorithm are promising showing good data through put with constant control packet overhead and it has been shown to perform well over a variety of mobility conditions. Reduced control traffic overhead, and quick route discovery and repair are the benefits of the termite algorithm. However, the termite algorithm has many tunable parameters and the efficiency of the algorithm depends on the values chosen for these parameters. Important among them is the pheromone update and decay which should reflect the current context of the network. Ruth [44] proposed many techniques to reflect the current context of the network in the termite routing algorithm and the first is the γ pheromone filter (γPF) or continuous pheromone decay. In the γ pheromone filter, whenever a node receives a packet, the pheromone on its entire neighbour links will be simultaneously decayed proportionally to the inter packet arrival time; further pheromone is also decayed when it is checked to send the packets. Each packet contains the pheromone γ equal to the utility of the path it has traversed and the γ pheromone filter is described by a pheromone update as shown in Equations (3.4) and (3.5).

$$\forall_i \quad P_{i,s}^n = P_{i,s}^n . e^{-(t-t_{s,obs}^n)\tau} \tag{3.4}$$

$$P_{r,s}^n = P_{r,s}^n + \gamma \tag{3.5}$$

where $P_{i,s}^n$ is the amount of pheromone from source node s, on the neighbor node i at node n; r represents the previous hop of the packet and γ is the amount of pheromone carried by each packet; τ is the pheromone decay rate; $t_{s,obs}^n$ is the last instant of time the pheromone is observed from source node s at node n whereas t gives the current time. Apart from the γ pheromone filter, the authors also proposed several other pheromone update and decay methods such as averaging filter, normalized γ pheromone filter (NγPF), probabilistic bellman-Ford (pBF), ant-based control + X, box filter and Oracle to keep the fresh routing information in the pheromone table. Ant-based control + X is a combination of ABC pheromone update methods with the γPF (ABC+γPF), NγPF (ABC+NγPF) and pBF (ABC+pBF).

Roth and Wicker [43] presented an analytical model to study the proposed pheromone update methods of the termite algorithm. The asymptotic behaviour of the algorithm is studied by using an analytical model and the time average pheromone value deposited over both single and double links is determined. The properties of the termite algorithm were investigated for the γ

pheromone filter, joint decay infinite impulse response filter-2 (IIR2) and pDijkstra pheromone update methods and the authors also explored the relationship between the parameters (decay rate and packet arrival rate) and found the scale invariance parameter. Through theoretical results, the authors have shown that the amount of pheromone on a link influences the performance of the routing algorithm in terms of both data through put and adaptability. The authors in [42] performed an analytical study on undesirable property of the termite routing algorithm in terms of tendency to converge on only one path between source and destination; thus the termite algorithm will not take full advantage of multipath routes to the destination. The authors used continuous pheromone decay function which works as follows. Whenever a node receives a packet, the pheromones over all its neighbor links will be decayed proportionally based on the interpacket arrival time and the link on which the packet has arrived is positively reinforced by updating the pheromone by γ. Continuous pheromone decay function is defined by

$$P^{n'}_{i,s} = \begin{cases} P^n_{i,s} \cdot e^{-(t-t^n_s)\tau} + \gamma & i = k \\ P^n_{i,s} \cdot e^{-(t-t^n_s)\tau} & i \neq k \end{cases} \tag{3.6}$$

where $P^{n'}_{i,s}$ is the amount of pheromone from source node s from neighbor node i at node n; k stands for the previous hop of the packet; γ is the amount of pheromone carried by the packet; τ is the decay rate and t^n_s is the previous time that a packet has arrived at node n from s whereas t gives the current time. The authors employed the normalized uncoupled pheromone update function to overcome the undesirable stagnation problem of the algorithm and thus distribute the traffic proportionally across the outgoing links. Pheromones on each link will be decayed only when a packet actually arrives on it and for normalization authors used the one tap IIR averaging filter and the box filter.

Ruth [44] proposed the ReTermite algorithm for MANETs which uses source pheromone repel (SPR) to push the packet away from its source and at the same time the packet will be pulled toward its destination. Source and destination pheromone distributions of ReTermite are calculated using Equations (3.7) and (3.8). The next hop is chosen randomly according to the meta distribution $\hat{p}^n_{i,d}$ as shown in Equation (3.9).

$$p^n_{i,d} = \frac{(P^n_{i,d} + K)^F}{\sum_{j \in N^n} (P^n_{j,d} + K)^F} \tag{3.7}$$

$$p^n_{i,s} = \frac{(P^n_{i,s} + K)^F}{\sum_{j \in N^n} (P^n_{j,s} + K)^F} \tag{3.8}$$

$$\hat{p}^n_{i,d} = \frac{p^n_{i,d} (p^n_{i,s})^{-R}}{\sum_{j \in N^n} p^n_{j,d} (p^n_{j,s})^{-R}} \tag{3.9}$$

The salient features of the ReTermite algorithm such as true continuous pheromone decay, piggybacked routing information and promiscuity were found to be very effective. The simulation results demonstrated that ReTermite outperforms the traditional AODV routing protocol.

3.3 Non-SI Based Bio-Inspired Routing Algorithms

Other animal societies such as mammal bats have also motivated several researchers for designing adaptive bio-inspired algorithms for solving complex real world problems. However, bats-based bio-inspired routing algorithms have not been found in the literature for networking and telecommunication applications but were found for optimizing non-networking problems such as brushless DC wheel motor problems, continuous constrained optimization problems [1] [49] [53].

3.4 Hybrid Bio-Inspired Routing Algorithms

The above discussed social insects and other animal societies have wonderful characteristics and indicate that it may be possible to combine the unique features of both social insects and animal societies for the design and development of novel hybrid routing algorithms for MANETs. The state-of-the-art hybrid bio-inspired routing protocol for MANETs is discussed.

Nature-Inspired Scalable Routing Protocol (NISR) (2011)
Jahanbakhsh et al. [17] proposed a hybrid nature-inspired scalable routing protocol for MANETs. It is known as NISR and includes the salient features of both ant and bee colonies. The protocol is derived from TORA. Scout bees are responsible for finding food sources and determining quality. When a scout bee returns to the hive, it indicates the distance to, direction of, quality, and quantity of the food source to the other bees by dancing. Based on information from the scout, the hive sends bees to the food source. When they reach the source, an ant and a bee are sent back to the hive. The ant updates the pheromone over the path and the bee informs the hive of the quality of the food source. Scout bees continuously search for new paths to new food sources to report to the hive.

Tables 3.1 through 3.9 present analyses of the bio-inspired routing protocols for MANETs discussed in this chapter. Figure 3.6 shows the taxonomy of state-of-the-art ACObased routing protocols and their properties.

TABLE 3.1
Congestion Aware ACO-Based Bio-Inspired Routing Algorithms

Algorithm	Parameters	Agents	Purpose of Agents	Remarks
ANSI	Distance to the destination in hops	F, B	Route discovery and setup	Targets congestion
EARA	Congestion	LFA		ARA with multipath, targets congestion
AntHocNet	Number of hops, congestion, signal-to-noise ration	F, B		Targets congestion, open space evaluation
ARAAI	Link stability based on node mobility	F, B		Targets stagnation
PAR	Queue length	F, B		Congestion aware, more exploratory

TABLE 3.2
Load and Mobility Aware ACO-Based Bio-Inspired Routing Algorithms

Algorithm	Parameters	Agents	Purpose of Agents	Remarks
MARA	Constant pheromone	F, B	Route discovery and setup	Load balancing
AntOR	Number of hops	F, B	Route discovery and setup	Load balancing
PACONET	Active connection time	F, B	r Route discovery and setup	Targets mobility and route maintenance

TABLE 3.3
CLD- and Hybrid-Based Bio-Inspired Routing Algorithm

Algorithm	Parameters	Agents	Purpose of Agents	Remarks
EARA-QoS	Delay, queue length	LFA	Route discovery and setup	CLD, QoS aware, targets diffServ
NISR	Constant pheromone	Ants, bees	Scout bees for continuous route exploration and setup; ants for reinforcement	Hybrid routing

TABLE 3.4
Location Aware ACO-Based Bio-Inspired Routing Algorithms

Algorithm	Parameters	Agents	Purpose of Agents	Remarks
GPSAL	No pheromone concept	F, B	Spreading routing information	Bandwidth efficient with flooding ants
MABR	No pheromone concept	F	Finding new paths and updating routing information	Part of AMRA; better than GFP and GPRS
HOPNET	Time taken to reach node K from source S	B, IFA, EFA, NA, EA	IFA maintains intrazone routing table; EFA maintains interzone routing table; B is for route setup; NA and EA handle route maintenance	Good for large networks with low and high mobility conditions
HRAZHLS	Time taken to reach node K from source S	B, IFA,EFA, NA, EA	IFA maintains intrazone routing table; EFA maintains interzone routing table; B is for route setup; NA and EA handle route maintenance	Targets routing overhead
ACAD	Anti-pheromone	F,B	Cluster detection	Automatic and shape-independent clustering

TABLE 3.5
Bees-Based Routing Algorithms

Algorithm	Parameters	Agents	Purpose of Agents	Remarks
BeeAdHoc	Energy	Scouts and foragers	Route establishment and maintenance	-
MABR	Energy	Scouts and foragers	Route establishment and maintenance	Targets security

TABLE 3.6
QoS Aware ACO-Based Bio-Inspired Routing Algorithms

Algorithm	Parameters	Agents	Purpose of Agents	Remarks
ADRA	Node velocity, queuing, delay, hop count	F, B, EA, AA	Route discovery and setup (F,B), congestion handling (EA, AA)	QoS aware, congestion and load balancing
ARAMA	Energy, processing power, link bandwidth	F, B, N, D	Route discovery and setup	QoS aware, energy efficient, fair resource distribution
SDVR	Delay, jitter, energy	F, B	Route discovery and setup	QoS aware, higher overhead
FACO	Energy, packet buffered in queue, signal strength	F, B	Route discovery and setup	*Fuzzy* based, QoS aware
ADSR	Delay, jitter, energy	F, B	Route discovery and setup	QoS aware
MQAMR	Bandwidth, delay, hop count, mobility, energy	F, B	Route discovery and setup	QoS aware, fair bandwidth allocation, longer network lifetime
S. Kim	Node distance, queue length, bandwidth, delay, energy	F, B	Route discovery and setup	Load balancing, QoS aware

TABLE 3.7
Termite-Based Routing Algorithms

Algorithm	Parameters	Remarks
Termite	Constant pheromone	Robust, reduces control packets.
Termite	Interpacket arrival time	Reflects current context of network
ReTermite	Interpacket arrival time	True continuous pheromone decay, piggybacked routing information promiscuity

TABLE 3.8
Other ACO-Based Bio-Inspired Routing Algorithms

Algorithm	Parameters	Agents	Purpose of Agents	Remarks
AAR	Queue length	F, B	Continuous route exploration and path setup	Targets convergence and average packet latency
ARA	Constant pheromone	F, B	Route discovery and setup	Targets Routing Overhead
AntAODV	No pheromone	RA	Route discovery and setup	Increased connectivity
PERA	Delay or number of hops	F, B	Route discovery and setup	Low throughput under high mobility condition
ABC-AdHoc	Constant pheromone	F	Continuous route exploration and path setup	Combination of ABC and AntNet
W_AntNet	Trip time, node queue status	F, B	Continuous route exploration and path setup	Good for static networks
SNP	Hop distance	RA	Spread destination scent odor	Localization and tracking
SARA	Constant pheromone	F, B	Route discovery and setup	Targets control packet overhead

TABLE 3.9
Flocks of Birds-Based Routing Algorithms

Algorithm	Parameters	Remarks
Tiwari and Varma	Energy	Scalable and cluster based
BFBR	Energy	Low route discovery latency and effective route maintenance

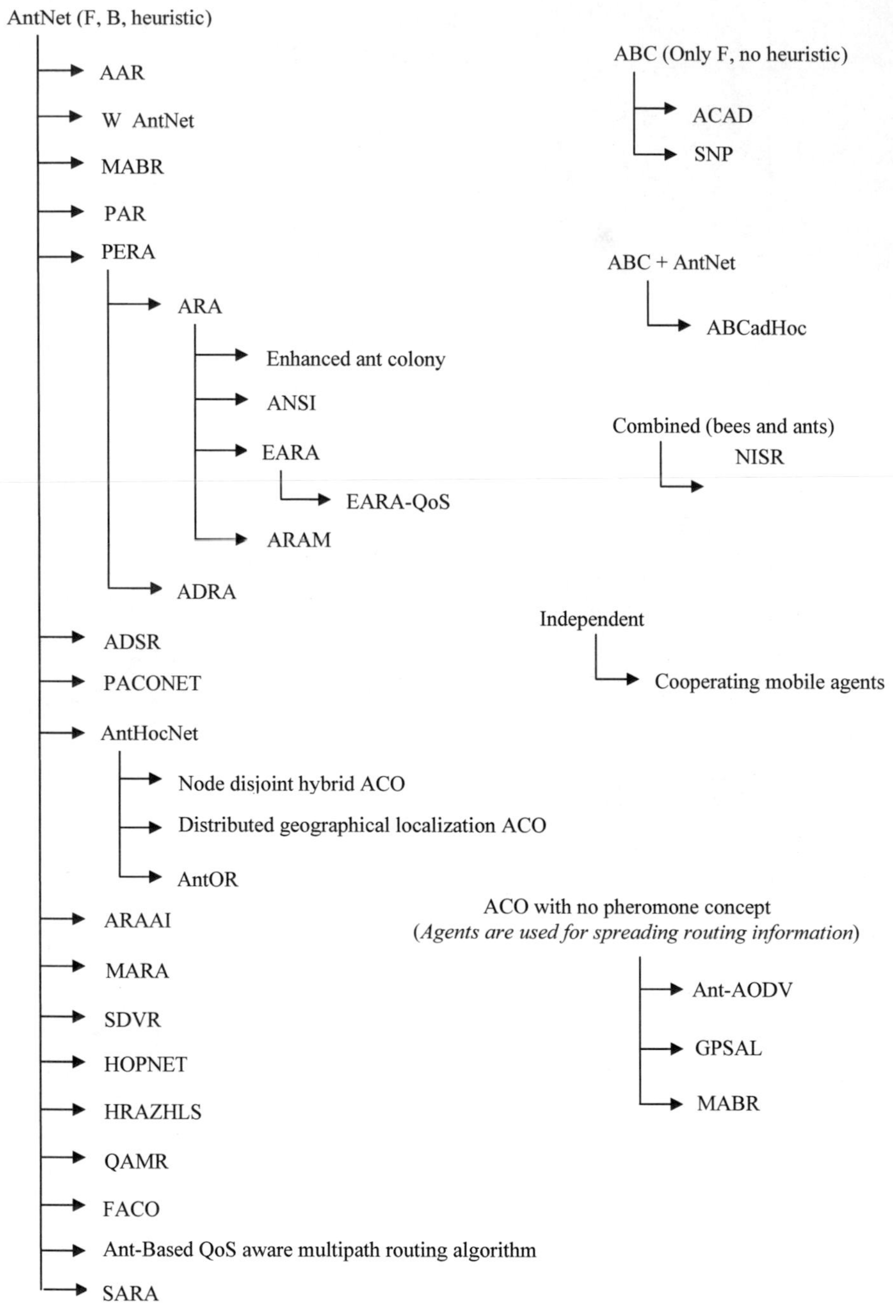

FIGURE 3.4
ACO Taxonomy based on Agents and Heuristics used in Routing

3.5 Summary

In this chapter, state-of-the-art bio-inspired routing protocols are discussed in detail along with several important features of social insects and animal societies. A detailed classification of SI-and non-SI-based routing algorithms is also given based on practices and properties and finally open issues and research challenges in designing routing algorithms for MANETs are highlighted.

Exercises

Part A Questions

1. The protocol in which route updates are propagated throughout a network at regular intervals to maintain the current topology information in each node is
   (a) Reactive
   (b) Proactive
   (c) Hybrid
   (d) None of the above
2. The optimized link state routing (OLSR) protocol is
   (a) Proactive
   (b) Reactive
   (c) Hybrid
   (d) None of the above
3. A reactive routing protocol used in MANETs is
   (a) ZRP
   (b) DSDV
   (c) OLSR
   (d) AODV
4. If a node in AODV is the intended destination, the packet sent by the node is
   (a) RERQ
   (b) RERR
   (c) RREP
   (d) None of the above

5. The list of nodes through which the packet is to be forwarded to reach the destination node in DSR is called the
   (a) Packet record
   (b) Route record
   (c) Host record
   (d) Route discovery
6. ACO-based routing algorithms for both wired and wireless communication networks are
   (a) ABC
   (b) AntNet
   (c) ACR
   (d) All of the above
7. A congestion aware ACO routing algorithm is
   (a) ANSI
   (b) EARA
   (c) AntHocNet
   (d) All of the above
8. The algorithm which uses forward and backward ants for route discovery and setup is
   (a) ARAAI
   (b) EARA
   (c) AntHocNet
   (d) None of the above
9. The algorithm designed to improve performance metrics such as delay and congestion and provide the best quality service using CLD across MAC and network layers is
   (a) EARA
   (b) MARA
   (c) AntOR
   (d) EARA-QoS
10. An example of load aware ACO routing protocols is
   (a) MARA
   (b) EARA
   (c) AntOR
   (d) Both (a) and (c)
11. The algorithm that uses GPS in location aware ACO routing protocols is

(a) GPSAL
(b) MABR
(c) HOPNET
(d) None of the above

12. The algorithm that divides the network area into rectangular zones and uses a logical router for long distance routing is

    (a) HOPNET
    (b) MABR
    (c) AntOR
    (d) MARA

13. The algorithm that combines ACO principles with zone routing protocol (ZRP) and DSR is

    (a) HOPNET
    (b) MABR
    (c) AntOR
    (d) MARA

14. The algorithm consisting of ants that randomly traverse the network and keep the record of last N visited nodes and these records updates in each visited node is

    (a) ARA
    (b) Ant-AODV
    (c) HOPNET
    (d) None of the above

15. An ACO-based routing algorithm that floods the forward ants only during the beginning of a communication session or when routes are stale is

    (a) AntNet
    (b) PERA
    (c) ABC
    (d) EARA

16. The protocol that gives scent-based guidance to ants for route discovery and maintenance is

    (a) SNP
    (b) SARP
    (c) AntNET
    (d) None of the above

Part B Questions

1. Explain the working principle of the AODV routing protocol.
2. Explain how swarm intelligence principles help to solve the problems of MANETs.
3. Explain the general routing model of bee-based routing algorithms for MANETs.
4. Explain the general routing model of flocks of birds-based routing algorithms in MANETs.
5. Explain how the termite algorithm finds the optimal route between the source node and the destination node in MANETs. Also explain the pheromone update and decay models of the basic termite routing algorithm model.
6. Give detailed classifications of location aware SI-based routing algorithms for MANETs.

Part A Answers

1. b
2. a
3. d
4. c
5. b
6. c
7. d
8. a
9. d
10. d
11. a
12. b
13. a
14. b
15. b
16. a

References

[1] Matti Airas. Echolocation in bats. In *Proceedings of Spatial Sound Perception and Reproduction*, 2003.

[2] Chowdhury Aritra and Das Swagatam. Automatic shape independent clustering inspired by ant dynamics. *International Journal of Swarm and Evolutionary Computation*, pages 33–35, 2012.

[3] Kalavathi B and Duraiswamy K. Ant colony based node disjoint hybrid multi-path routing for mobile ad hoc networks. *Journal of Computer Science*, 4(2):80–86, 2008.

[4] B. Tatomir and L. Rothkrantz Dynamic routing in mobile wireless networks using ABC-AdHoc In *The Fourth International Workshop on ant Colony Optimization and Swarm Intelligence*, Springer, volume 3172, 2004.

[5] J.S. Baras and H. Mehta. A probabilistic emergent routing algorithm (PERA) for mobile ad hoc networks. In *Proceedings of Modeling and Optimization in Mobile, Ad Hoc and Wireless Networks*, 2003.

[6] D. Camara and A.F. Loureiro. A novel routing algorithm for ad hoc networks. In HICSS, *IEEE Press*, 2000.

[7] D. Camara and A.F. Loureiro. GPS/ant-like routing in ad hoc networks. *Telecommunication Systems*, 18(1–3):85–100, 2001.

[8] Gianni A. Di Caro, Frederick Ducatelle, and Luca M. Gambardella. *Theory and Practice of Ant Colony Optimization for Routing in dynamic Telecommunications Networks.* Idea Group, Hershey, Pa, 2008.

[9] Perkins C.E. and Royer E.M. Ad hoc on demand distance vector (AODV) routing. In *Second IEEE Mobile Computing Systems and Applications*, pages 90–100, 1999.

[10] Fernando Correia and Teresa Vazao. Simple ant routing algorithm. In *International Conference on Information Networking, IEEE*, pages 1–8, 2008.

[11] Fernando Correia and Teresa Vazao. Simple ant routing algorithm strategies for a (multipurpose) MANETs model. *Ad Hoc Networks,* 8:810–823, 2010.

[12] Sivakumar D and Bhuvaneshwaran RS. Proposal on multi agent-ants-based routing algorithm for mobile ad hoc networks. *International Journal of Computer Science and Network Security*, 17(6), 2007.

[13] Arbona X Dhillon S. and Van Mieghem P. Ant routing in mobile ad hoc networks. In *Proceddings of Third International Conference on Networking and Services, IEEE Computer Society*, pages 67–74, 2007.

[14] G. Di Caro and M. Dorigo. Mobile agents for adaptive routing. In *System Sciences, 1998., Proceedings of Thirty-First International Conference,* volume 7, pages 74–83, 1998.

[15] L.M Gambardella, F. Ducatelle and G A. Di Caro. Using ant agents to combine reactive and proactive strategies for routing in mobile ad hoc networks. *International Journal of Computational Intelligence and Applications*, 5(2):169–184, 2005.

[16] Muddassar Farooq and Gianni A. Di Caro. Routing protocols for next generation networks inspired by collective behaviors of insect societies: an overview. *Swarm Intelligence, Natural Computing Series*, pages 101–160, 2008.

[17] Jamali Shahram, Gudakahriz Sajjad Jahanbakhsh, and Zeinali Esmaeel. NISR: a nature inspired scalable routing protocol for mobile ad hoc networks. *International Journal of Computer Science Engineering and Technology*, 1(4):947–950, 2011.

[18] H. Matsuo and K. Mori Accelerated ant routing in dynamic networks. In *2nd ACIS International Conference on Software Engineering, Artificial Intelligence, Networking and Parallel and Distributed Computing*, pages 333–339, 2001.

[19] Zygmunt J. Haas and Marc R. Pearlman. The zone routing protocol (ZRP) for ad hoc networks. Technical report, Cornell University, 1997.

[20] Guoyou He. Destination-sequenced distance vector (DSDV) protocol. Technical report, Helsinki University of Technology, 2002.

[21] Jorg D Heissenbuttel, M Braun and T. Huber A framework for routing in larger ad hoc networks with irregular topologies. In *Challenges in Ad Hoc Networking, International Federation for Information Processing*, volume 197, pages 119–128, 2006.

[22] P. Jacquet, P. Muhlethaler, T. Clausen, A. Laouiti, A. Qayyum, and L. Viennot. Optimized link state routing protocol for ad hoc networks. In *Proceeding of Conference of Technology for the 21st Century. IEEE International*, pages 62–68, 2001.

[23] David B. Johnson, David A. Maltz, and Josh Broch. DSR: The dynamic source routing protocol for multi-hop wireless ad hoc networks. In *In Ad Hoc Networking,* Chapter 5, pages 139–172. Addison-Wesley, 2001.

[24] D.R. Canas L.J.G. Villalba and A.L.S. Orozco. Bio-inpired routing protocol for mobile ad hoc networks. *IET Communications*, 4(8):2187–2195, 2010.

[25] M. Kudelski and A. Pacut Ant routing with distributed geographical localization of knowledge in ad-hoc networks. In EvoCOMNET, LNCS, Springer, pages 111–116, 2009.

[26] U. Sorges, M. Gunes and I Bouazizi. ARA the ant-colony based routing algorithm for manets. In *International Workshop on Ad Hoc Networks*, pages 79–85, 2002.

[27] Eric Bonabeau Marco Dorigo and Guy Theraulaz. Ant algorithm and stigmergy. *Future Generation Computer Systems*, 16(8):851–71, 2000.

[28] Sanaz Asadinia, Marjan Kuchaki Rafsanjani and Farzaneh Pakzad. A hybrid routing algorithm based on ant colony and ZHLS routing protocol for MANETs. *Communications in Computer and Information Science*, 120:112–122, 2010.

[29] C. K. Tham, S. Marwaha, and S. Srinivasan. Mobile agent-based routing protocol for mobile ad hoc networks. In *Proceedings of IEEE Global Telecommunications Conference*, IEEE, Washington, DC, pages 163–167, 2002.

[30] N. Mazhar and M. Farooq. BeeAIS: artificial immune system security for nature-inspired MANET routing protocol BeeAdHoc. In *Proceedings of 6th International Conference on Artificial Immune Systems*, Springer, volume 4628, 2007.

[31] K. H. Kramer, N. Minar, and P. Maes. Cooperating mobile agents for dynamic network routing. In *Software Agents for Future Communication Systems*, Chap. 12. Springer, 1999.

[32] Mohit Tiwari and Shirshu Varma. Bird flight-inspired clustering-based routing protocol for mobile ad hoc networks. *International Journal of Computer Science and Network Security*, 10(3), 2010.

[33] C. Siva Ram Murthy and B. S. Manoj. *Ad Hoc Wireless Networks.* 2nd Edition, Pearson Education, 2005.

[34] Cauvery N and Vishwanatha K. Enhanced ant colony based-algorithm for routing in mobile ad hoc networks. In *Proceedings of World Academy of Science, Engineering and Technology*, volume 36, pages 30–35, 2008.

[35] Thulasiraman Parimala, Osagie Eseosa, and K. Thulsiram Ruppa PACONET: ImProved ant colony optimization routing algorithm for mobile ad hoc networks. In *Proceedings of 22nd International Conference on Advanced Information Networking and Applications*, pages 204–211, 2008.

[36] Y.P Singh S. Prasad and CS. Rai. Swarm based intelligent routing for MANETS. *International Journal of Recent Trends in Engineering*, 2009.

[37] J. Bruten, R. Schoonderwoerd, O. Holland, and L. Rothkrantz. Ant-based load balancing in telecommunication networks. HP Labs Technical Reports, 1996.

[38] M.A. Rahman, F. Anwar, J. Naeem, and M.S.M. Abedin. A simulation-based performance comparison of routing protocol on mobile ad hoc network (proactive, reactive and hybrid). In *Computer and Communication Engineering International Conference*, pages 1–5, 2010.

[39] Fred E. Regnier and John N. Insect pheromones. *Journal of Lipid Research*, 14, 1968.

[40] Martin Roth and Stephen Wicker. Termite: ad hoc networking with stigmergy. In *IEEE Global Communication Conference*, volume 5, pages 2937–2941, 2003.

[41] Martin Roth and Stephen Wicker. Termite: Emergent ad hoc networking. In *2nd Mediterranean Workshop on Ad Hoc Networking*, 2003.

[42] Martin Roth and Stephen Wicker. Asymptotic pheromone behavior in swarm intelligent MANETs: an analytical analysis of routing behavior. In *Sixth IFIP IEEE International Conference on Mobile and Wireless Communication Networks*, 2004.

[43] Martin Roth and Stephen Wicker. Network routing with filters: link utility estimation in swarm intelligent manets. Technical report Defence Advance Research Project Agency (DARPA) and Army Research Office, 2004.

[44] Martin Heinz Ruth. Termite: A Swarm Intelligent Routing Algorithm for Mobile Ad Hoc Networks. PhD thesis, Cornell University, 2005.

[45] S. Rajagopalan and C. Shen ANSI: a unicast routing protocol for mobile ad hoc networks using swarm intelligence. In *proceedings of International Conference on Artificial Intelligence*, pages 104–110, 2006.

[46] H.F. Wedde and M. Farooq. An energy aware scheduling and routing framework. School of Computer Science, University of Dortmund.

[47] Cauvery. N.K, G.S Sharvani and T.M. Rangaswamy. Different types of swarm intelligence algorithm for routing. In *International Conference on Advances in Recent Technologies in Communication and Computing*, pages 1–5, 2009.

[48] T. Niranjan, B. T Srinivasan, V. Mahadevan et al. BFBR: a novel bird flocking behavior-based routing for highly mobile ad hoc networks. In *Proceedings of International Conference on Computational Intelligence for Modelling Control and Automation*, IEEE, 2006.

[49] Leandro dos S. Coelho, Teodoro C. Bora, and Luiz Lebensztajn. Bat-inspired optimization approach for the brushless dc wheel motor problem. *IEEE Transactions on Magnetics*, 48(2):947–950, 2012.

[50] Osagie Eseosa, Wang Jianping, Thulasiraman Parimala, and K. Thulasiram ruppa, HOPNET: a hybrid ant colony optimization routing algorithm for mobile ad hoc networks. *Ad Hoc Networks*, 7:690–705, 2009.

[51] T. D. Wyatt. *Pheromones and Animal Behaviour: Communication by Smell and Taste.* Cambridge University Press, 2003.

[52] Song Luo, Yalin Evren Sagduyu, and Jason H. Li. On the overhead and throughput performance of scent-based MANET routing. In *The Military Communications Conference on Unclassified Program Networking Protocols and Performance Track, IEEE*, 2010.

[53] Xin-She Yang. A new metaheurastic bat-inspired algorithm. In *Nature Inspired Cooperative Strategies for Optimization, Studies in Computational Intelligence, Springer*, 2010.

[54] Yuan Yuan Zeng and Yan Xiang. Ant routing algorithm for mobile ad hoc networks based on adaptive improvement. In *Proceedings of International Conference on Wireless Communications, Networking and Mobile Computing*, volume 2, pages 678–681, 2005.

[55] Marta Z. Kwiatkowska, Zhenyu Liu, and Costas Constantinou. A biologically inspired QoS routing algorithm for mobile ad hoc networks. In *Proceedings of lASTED International Conference on Internet Communications and Information Technology*, 2004,

[56] Marta Z. Kwiatkowska, Zhenyu Liu, and Costas Constantinou. A biologically inspired QoS routing algorithm for mobile ad hoc networks. *In Proceedings of International Conference on Advanced Information Networking and Applications*, IEEE, pages 426-431, 2005.

4

SI Solutions to QoS in MANETs

CONTENTS

Providing QoS to end users is one of the most important challenges in today's telecommunication industry for supporting a wide range of real-time and non-real-time applications. QoS can be defined as a measure of service quality offered by a network to end users and applications [8]. The frequent link failures in MANETs due to the mobility of nodes makes source nodes spent most of their time in route setup and maintenance instead of sending messages. MANETs suffer from low throughput, frequent end-to-end delays, and high control packet overhead.

Finding stable nodes for routing is critical and challenging [15] [25] [33] because MANETs require context aware adaptive decision making environments under dynamic network conditions to provide QoS to applications.

Cross layer solutions for providing QoS for MANETs based on SI principles will be discussed in detail. Adaptive routing protocols that utilize only SI principles are also discussed. The parameter considered, the QoS parameter targeted, and the advantages and disadvantages of each protocol are examined in deep.

4.1 QoS Parameter at Different Layers

QoS defines parameters for traditional networks such as packet loss, delay, jitter, and bandwidth [32]. However, the QoS requirements in MANETs such as *fault tolerance, network lifetime, and control packet overhead*, are application specific and different from the traditional end-to-end QoS requirements. Although, QoS solutions such as IntServ, and DiffServ were developed for traditional networks, these cannot be easily ported to MANETs due to characteristics such as limited resources and capabilities of nodes, dynamic network topology, scalability, various types of applications, various traffic types, wireless link unreliability and data redundancy [11].

4.2 Providing QoS at Different Layers

All layers of a communication protocol stack aim to provide desired QoS levels to the application end users while decreasing protocol overhead and consequently prolonging network lifetime. QoS provisioning in MANETs can be accomplished by considering three layers of protocol stacks: MAC, network and transport layers.

1. The MAC layer aims at providing QoS and minimizing medium access delay, collisions, energy consumption by interference and maximizing reliability and concurrency (parallel transmissions).
2. The network layer tries to minimize path latency and energy consumption and maximize routing reliability, and path lifetime to provide QoS to the application end user.
3. The transport layer tries to minimize energy consumption, and minimize packet loss and maximize end-to-end reliability, and provide bandwidth and throughput fairness, thereby ensuring QoS to the end user applications.

The transport layer in MANETs is also responsible for avoiding possible congestion and delivering data reliably to the sink in the network to improve QoS to the applications and end users [28]. Congestion leads to packet drops that may delay important information and it also leads to waste of both energy and time of a node when reprocessing dropped packets. Congestion in

MANETs must be solved efficiently to avoid energy waste and increase the lifetime of a network, thereby increasing its throughput.

4.3 SI- and CLD-Based Solutions to QoS

In recent years, a lot of research has been carried out to develop SI-based adaptive bioinspired routing protocols for MANETs. Because centralized approaches for routing in MANETs lack scalability and fault tolerance, SI techniques yield natural solutions through a distributed approach to adaptive routing. Another promising method called cross layer design (CLD) can also be used to achieve QoS in MANETs.

Because of dependencies between layers in protocol stacks, QoS metrics such as maximizing overall throughput, minimizing end-to-end delay and minimizing congestion probability are difficult to achieve with a layered QoS approach. Individual layers cannot guarantee overall QoS provisioning and a cross layer solution arises as a promising method which provides more efficiency by sharing important information among all layers of a protocol stack. CLD could be used for optimizing OSI layer performance, thereby of providing QoS support to applications and end users of MANETs. Primary QoS metrics that may be met with a cross layer approach are (1) minimizing overall energy consumption; (2) minimizing end-to-end delays; (3) maximizing network lifetimes; (4) maximizing network throughputs; and (5) maximizing overall reliability. CLD can also be used to improve other QoS metrics such as minimizing collisions and congestion and providing effective sampling rates [14].

Thus both SI and CLD techniques are promising techniques for providing expected levels of QoS to application end users of MANETs. Further, by combining the distinguishing features of both SI and CLD, very high levels of QoS can also be provided to end users. Hence in this chapter, the routing protocols which combine the features of both SI and CLD to provide QoS to the application end users are discussed in detail.

4.4 Taxonomy of SI-Based QoS Protocols

In this section the state of the art of bio-inspired adaptive routing protocols for MANETs that display the salient features of CLD to provide QoS to end users are discussed in detail. The bio-inspired routing protocols such as mobility aware termite (MA-termite) [19] and load-balanced termite (LB-T) [18] which are based on the behavior of termites will be discussed first. Next, the bio-inspired hybrid routing protocols for MANETs combining the features of both termites and bats known as bat-termite [20] and load-balanced bat-termite (LB-bat-termite [21] will be described. The next section covers the working principles of routing protocols for MANETs based on the characteristics of a single insect (ant, termite, or bee) in their designs.

4.4.1 Cross Layered-SI based QoS Protocols

The mobile nodes found in existing MANETs are capable of monitoring network status as well as data processing during an ongoing data session. Thus, context aware routing in MANETs can be achieved by local monitoring capability of nodes. The context awareness helps the mobile nodes know the current status of their neighbors such as (mobility, available bandwidth, queue status), thereby deciding whether to include the perticular neighbor in the path or not. The context awareness also helps the mobile nodes in achieving the desired QoS by choosing the resource-rich and reliable nodes in the path. Thus in this section, the context aware bio-inspired CLD-based routing protocols for MANETs which target QoS are discussed in detail.

MA-termite [19] is a novel bio-inspired routing protocol for MANETs based on the behavior of termites that tries to add context awareness (current status of neighboring nodes) in all nodes. The primary objectives of MA-termite are (1) positively reinforcing stable nodes to reduce pheromone decay and (2) impacting unstable nodes by encouraging high levels of pheromone decay. A secondary objective is providing QoS to end users of MANETs by ensuring reliable paths between source and destination nodes. MA-termite defines distance-based pheromone and decay functions to reflect the current state of node mobility in a network. The pheromone update and decay over the links are directly proportional to the distances between nodes on successive packet arrival, thus causing high pheromone decay in high mobility nodes and low pheromone decay in low mobility nodes. During data transfer, low mobility nodes will be given preference over high mobility nodes. The coordinated behavior exhibited by termites during the hill building phase in MA-termite is used to guide packets to reach their destinations. Each node is treated as a termite hill that contains a pheromone table for storing the pheromone gradients of all reachable destinations. Packets leaving the source node are forwarded toward the neighbor node with the highest destination pheromone gradient and the selection of the next hop is decided probabilistically at every node. For that reason MA-termite is known as a probabilistic routing algorithm.

At a given node, the packet arrival updates the source pheromone table proportional to the efficiency of the path traversed. MA-termite removes the old routing solutions by mimicking the pheromone decay concept followed by termites. MA-termite is driven mainly by pheromone update, decay, and forwarding functions. If a node must send a data packet and the destination pheromone entry is not present in its table, the initial node will first buffer the data packet and then trigger a route set up (route discovery) phase in which a new pheromone trail to the destination is found. After the new route (pheromone trail) to the destination is found, MA-termite forwards all buffered data packets and initiates data session phases between the source and destination nodes.

The proposed algorithm exhibits salient features such as high robustness through mUltiple paths to destinations, sufficient route reliability with context awareness, and fewer control packets with stable nodes in the paths. Each node in MA-termite will have a general idea about the positions of neighboring nodes and their movements and this will enable it to find stable nodes en route to its destination.

In any ant- or termite-based routing protocols, the pheromone update and decay functions will play very important roles. A slow pheromone decay rate will cause stale entries in the pheromone table, thus resulting in wrong decisions by the nodes and a high decay rate may quickly remove the pheromone entry from the pheromone table, thus closing the alternate path to the destination. To choose the optimal decay rate, the interpacket distance between the nodes is considered for pheromone update and decay [12] in each node. The interpacket distance between the nodes reflects the mobility of each neighbor node as shown in Figure 4.1, thereby adding context awareness in the each mobile node in the network.

If the interpacket distance between the nodes is small (Figure 4.1a), the decay will also be small. On the other hand, the decay will be large when the interpacket distance between the nodes is large (Figure 4.1b), thus reflecting the current context of the network. Accordingly the pheromone update and decay function is designed in such a manner that it meets the requirements of MA-termite by finding the distance between the nodes using the cross layer design model as explained below.

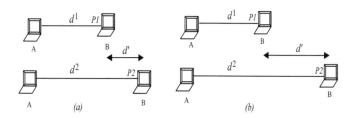

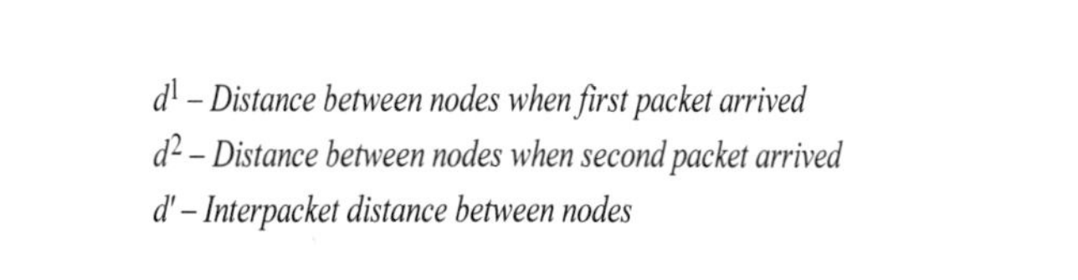

FIGURE 4.1
Interpacket Distance between Nodes

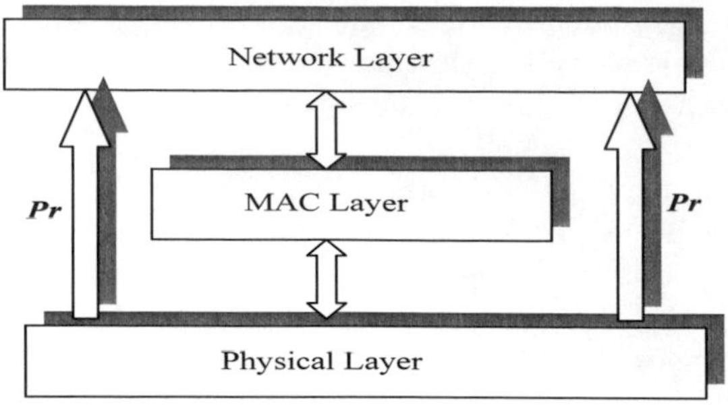

FIGURE 4.2
Cross Layer Design Model

Cross Layer Design Model
Figure 4.2 shows the cross layer design model across the physical and network layers which can be used for finding the distance between the nodes.

Since received signal strength *(Pr)* is inversely proportional to the distance between nodes [17] [1], variable *Pr* of each packet from the physical layer is made visible at the network layer in MA-termite. Whenever any mobile node receives a packet from a neighbor node, it first measures the *Pr* and then calculates the distance d to the neighboring node using the free space propagation model as shown in Equation (4.1). The computed distance d and the *time interval* (which gives the freshness of the data) at which d is computed will be stored in corresponding table of the network layer. Table entries are refreshed at regular intervals to remove the stale entries. The computed distance d will be used for updating and decaying the trail pheromone in mobile nodes and the details of pheromone update and decay functions are given in the following subsection.

Trail Pheromone Update-Decay Function
The continuous pheromone update and decay function of the basic termite algorithm [27] [26] has been modified to suit the requirements of the proposed MA-termite algorithm and is as follows.

$$P_R^n = \begin{cases} P_{k,s}^n \cdot e^{-(d_c - d_p)\tau + \alpha} & k = l \\ P_{k,s}^n \cdot e^{-(d_c - d_p)\tau} & k \neq l \end{cases} \tag{4.1}$$

where P_R^n is the amount of pheromone from source node s deposited on neighbor link k at node n; l represents the previous hop of the packet; α is the amount of pheromone carried by the packet; d_c represents the current distance of the neighbor node k; d_p is the distance of the same neighbor node k when the last packet arrived; and τ is the decay rate.

When a packet arrives at a mobile node, the pheromones on its neighbor node links will be decayed simultaneously based on the distance travelled by the neighbor nodes. Pheromone update is handled solely by the neighbor link from which the data packet arrived and equals the amount of pheromone carried by the data packet, thereby reinforcing the corresponding neighbor link positively. The computed trail pheromone of each neighbor node will be used by the forwarding function for finding the probability of a neighbor node reaching the destination and thus forwarding the data packets toward the destination node. The next section describes details of the forwarding function.

Forwarding Function of MA-termite Algorithm

The forwarding function [27] [26] handles each data packet independently and it is given as

$$p^n_{i,d} = \frac{(P^n_{i,d} + K)^F}{\sum_{j=1}^{N}(P^n_{j,d} + K)^F} \tag{4.2}$$

where $p^n_{i,d}$ is the probability of using neighbor node i to reach destination d at node n and N is the total number of neighbor nodes of$n$; constants K and F indicate the *pheromone threshold* and *pheromone sensitivity* used for tuning the routing behavior of MA-termite, respectively.

To store the computed trail pheromone data and current status of neighbor nodes, MAtermite utilizes tables in the network (the local status table and trail pheromone table). The node structure of MA-termite is explained below.

Node Structure

Figure 4.3 illustrates the node structure of the MA-termite algorithm. Each node consists of two tables. The forwarding function uses the trail pheromone table (TPT) to decide the next node along the path. The update function uses the local status table (LST) to update the pheromone on the links. The node structure is shown in Figure 4.3.

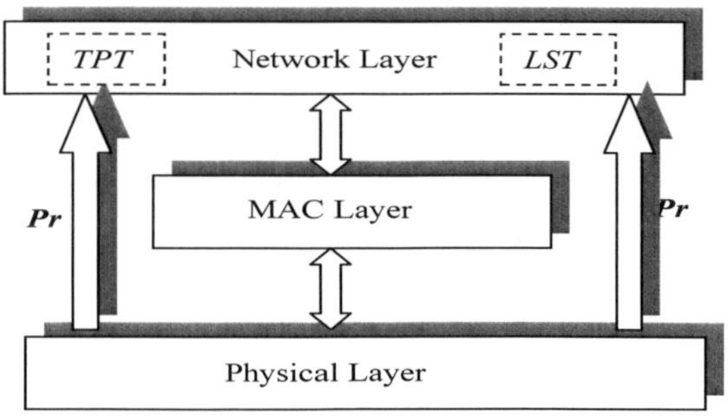

FIGURE 4.3
Node Structure of MA-Termite Algorithm

Pheromone Bounds

MA-termite uses three bounds for the pheromones; PH_BEGIN, PH_MIN and PH_MAX to limit the pheromone value in the pheromone table. Whenever a node receives a packet from an unknown source, a new entry will be created for that node in the TPT with an initial value of PH_BEGIN. Pheromones in the pheromone table will not be allowed to fall below the minimum value PH_MIN exceed the maximum value PH_MAX and these pheromone bounds will be chosen according to the context of the network. The details of trail pheromone table are given below.

Trail Pheromone Table

The TPT records the amount of destination pheromone on each neighbor link; TPT helps establish the best path from the source to the destination and more or less holds the same elements as a routing table in terms of neighbor identification, destination identification, and routing metric (trail pheromone in case ofMA-termite). Rows represent neighbor nodes and columns correspond to destination nodes. When a node detects a new neighbor node, it adds an additional row to the TPT. When a new destination is discovered, a new column is added. In both cases, the trail pheromone is initialized to PH_BEGIN. TPT is updated based on the local status table (LST) ofthe node as described below.

Local Status Table

The LST indicates the current context of the network in terms of the neighbor nodes' distance information which is maintained by the network layer and the TPT. The LST shows neighbor nodes' current and previous distances between the successive packet arrivals and the times at which the distances were calculated. The time entries maintain the freshness of LST information; the LST is refreshed at regular intervals of LST_TIME_OUT.

Whenever a node receives a control (RREQ, RREP, or RERR) or data packet, it will calculate the distance of the transmitting node using Equation (4.1) and store it in the LST. Thus every node is aware of its neighboring node movements and the chances that neighbor nodes will stay within transmission range. Based on current and previous distances of neighboring node information from the LST, the trail pheromone will be updated in the TPT to reflect neighbor node movements.

Thus, a stable node has a higher pheromone entry in the TPT whereas an unstable neighbor node has a lower pheromone entry. LST and TPT are updated and maintained during the route discovery and maintenance phases of MA-termite and the details are given in the following subsection.

Route Setup and Maintenance Phase of MA-termite

When a node does not find the destination pheromone entry in its TPT in forwarding the data packet toward the destination, it first buffers the data packets and then broadcasts a route request (RREQ) packet toward its neighbor nodes. RREQ takes a random walk across the network and thus finds a new pheromone trail for the intended transmission. During the route setup phase, the RREQ packets will not be attracted by the highest pheromone link and will be forwarded to a random next hop other than the link on which it has arrived; if a node cannot forward the RREQ packet, it will automatically drop the RREQ packet. Whenever a node receives the RREQ packet, it makes the necessary entries in its TPTset up the reverse path and replies to the RREQ source with a route reply packet (RREP) only if (1) the node is the intended destination or (2) the node has the destination trail pheromone. The RREP will set up a forward path in each visited node. When the source receives the RREP, it forwards the buffered data packets according to the forwarding function and initiates a data session. The path chosen for the data session will be strengthened further by the data packets that traverse it.

During a data session, a node that cannot reach a neighbor node for forwarding data packets and a destination node that has no suitable neighbors will buffer the data packets and send the RREQ to find a new pheromone trail to the intended destination.

Algorithm 4.1 illustrates the route setup phase of MA-termite and Algorithm 4.2 describes the data session phase.

The proposed MA-termite algorithm is simulated in NS-2 and the performance results are compared with bio-inspired algorithms (SARA [7] and the basic termite algorithm [26] [27]) and non-bio-inspired algorithms (ad hoc on demand distance vector or AODV [5]). The protocols are compared with respect to QoS metrics such as throughput, total packet drops, end-to-end delays and control packet overhead. Two different test scenarios are considered to perform the qualitative and quantitative analyses under scalability and node velocity conditions of the MA-termite algorithm.

Scenario A Scenario scalability of the proposed algorithms is evaluated by considering two test areas analogous to of size 500 m × 500 m and 1000 m × 1000 m. Under each simulation area, the algorithms are tested for density of 100, 200, 400 and 600 nodes. The node velocity is kept constant at 2m/s when a normal walking scenario of a person is considered and no pause time is given to any mobile node.

Scenario B Behaviors of the proposed algorithms under different sets of node velocity are studied for performance evaluation. Mobile nodes with varying velocities (2, 4, 6, 8 and 10 m/s) and different densities (100, 200, 400 and 600 nodes) are considered for each simulation run. Test area analogous to size 1000 m × 1000 m is considered and no pause time is given to any mobile node.

From the simulation results of MA-termite, we can see that it achieves 2 to 3% improvement when compared to SARA with respect to QoS metrics

Algorithm 4.1: Route Setup Phase of MA-termite

Input: TPT and LST

Output: Updated TPT and LST and optimal path between source and destination node.

1. ***If*** node n does not find p_d^n in its TPT to forward data packets,
2. Buffer data packets
3. Release *RREQ* packets from node n to find new pheromone trail to destination

1. ***When*** an intermediate node receives *RREQ* packet,
2. ***if*** it is destination or consists of pheromone trail to destination
3. Unicast *RREP* packet to source
4. Discard *RREQ*
5. ***else***
6. Update TPT and LST
7. ***if*** neighbor exists
8. Randomly choose neighbor and forward RREQ packet
9. ***else***
10. Discard RREQ
11. ***End if***
12. ***End if***
13. ***End When***

1. ***If*** node n receives *RREP* packet within *RREQ_TIME_OUT*,
2. Forward buffered data packets and start data session
3. ***Else***
4. Send *RREQ* packets from node n to find new pheromone trail to destination
5. ***End if***

such as throughput, end-to-end delay and total packet drops. MA-termite also achieves 5 to 8% improvement when compared to basic termite and AODV algorithms in terms of QoS metrics such as *throughput, end-to-end delays, total packet drops* and *control packet overhead.* Simulation results also demonstrate that MA-termite exhibits superior performance when compared to SARA, basic termite and AODV under dynamic network conditions.

The asymptotic pheromone behavior of MA-termite reveals that it always chooses the highest pheromone link at any given time and thus suffers from stagnation. It MA-termite also suffers from poor network exploration and improper backup path maintenance which cause the early removal of secondary paths from the TPT. Thus, in the next section a hybrid bio-inspired routing protocol based on the behaviors of termites and bats will be discussed to overcome the poor network exploration and improper backup path maintenance problems of MA-termite.

Algorithm 4.2 Data Session Phase of MA-termite

Input: TPT and LST

Output: Updated TPS and LST and reinforcement of optimal path between source and destination.

1. ***while*** data packet reaches destination node
2. At node x which is intermediate node in path
3. ***for*** all neighbors of node x
4. Decay pheromone
5. ***end for***
6. for neighbor y from which data packet arrived
7. Update $p^x_{s,d}$ in x's TPT where s is source node
8. Update path utility value in data packet
9. Calculate next-hop probability for all neighbors of x which lead to destination
10. Choose highest probability neighbor and forward data packet
11. ***if*** no neighbor exists for x that leads to the destination
12. Buffer data packet
13. Initiate route setup phase
14. ***end if***
15. ***end while***

4.4.2 Bat-Termite: Novel Hybrid Bio-Inspired Routing

This section discusses a novel hybrid bio-inspired routing protocol for MANETs referred to as bat-termite [20] which combines the unique features of both termites and bats. The termite is well known for its hill building nature and bats are fascinating mammals known for their advanced and distinguishing echolocation capabilities. The bat-termite protocol combines the major behavioral advantages of termites and bats to build a new hybrid bio-inspired algorithm for MANETs. The echolocation feature will be utilized to solve the problems of poor network exploration and improper backup route maintenance of the MA-termite algorithm.

Motivation for Development of Bat-termite Algorithm

As discussed in the previous section, the MA-termite algorithm suffers from two drawbacks that weaken its performance: (1) poor network exploration and (2) improper backup route maintenance. Poor network exploration prevents MA-termite from adapting to the current context of a network; improper backup route maintenance causes MA-termite to neglect neighbor nodes to reach a destination even though they are active and reachable.

Asymptotic Behavior of MA-termite Algorithm

To demonstrate the asymptotic behavior of MA-termite, consider a two-path system with a source node X and a destination node Y as shown in Figure 4.4a. Node X has two autonomous paths to reach node Y through neighboring nodes P and Q as shown in Figure 4.4b. Since link *L2* is dominant and is

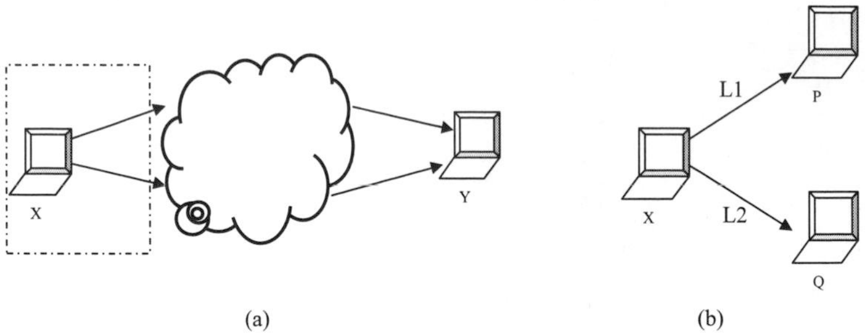

FIGURE 4.4
Asymptotic Behavior of MA-termite Algorithm

positively reinforced, the packets will choose *L2* in reaching the destination node *Y*. Eventually *L2* will attract the all traffic, thus accumulating more and more pheromone concentration on *L2*.

L1 will attract fewer packets, thereby spreading negative feedback and find less pheromone concentration on *L1*. The pheromone deposited on *L1* will be gradually decayed over certain period; under these circumstances, a single dominant link system will emerge from this double link system. Thus, MA-termite will always exclusively choose the dominant link *L2* and the link *L1*, which is less dominant, will gradually lose its priority over *L2*. Further, since the pheromone deposited on *L1* will reach the minimum value over time, the corresponding *L1* entry will be removed from the TPT of node *X*. MA-termite will close another option of reaching the destination node *Y* using link *L1*.

This example shows the poor backup root maintenance exhibited by MA-termite. Under such circumstances, even though the neighbor node *P* is reachable to the node *X* and could be used to reach the destination, *X* cannot use the facility of node *P* because its pheromone entry will not be present in node *X's* TPT. Under the worst case scenario, if *L2* fails for any reasons, node *X* will be left with empty neighbor node entries in its TPT even though a neighbor node *P* is active and could be used for reaching the destination. Thus, node *X* has to find an alternative pheromone trail to forward the data packets to the destination and in turn will create a hiccup in the data transmission. If node *X* has an option for checking the existence of node *P* and its current status before removing its pheromone entry from the TPT, node *X* could maintain an alternate route in reaching the destination under a worst case scenario and node *X* could have continued its data transmission using *L1* without hiccups. Thus, the bat-termite algorithm utilizes the fascinating echolocation feature of bats to address the improper backup route maintenance and poor network exploration of the MA-termite algorithm. The echolocation feature is explained in the next section.

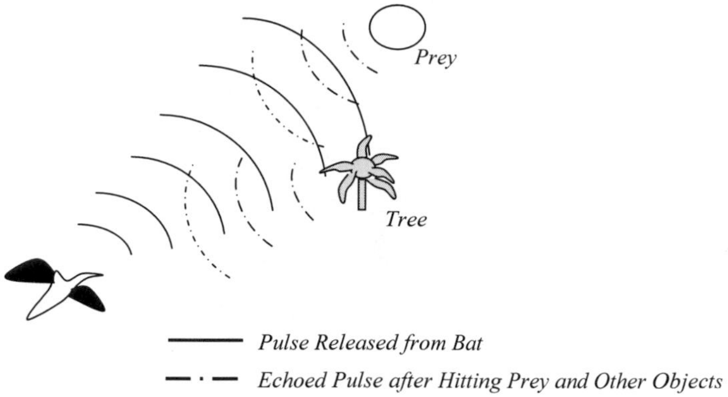

FIGURE 4.5
Echolocation in Bats

Echolocation in Bats
Two distinct characteristics of makes distinguish them from other mammals: (1) they have advanced echolocation capability and (2) unlike all other mammals, they have sings. Bats use their echolocation ability to find prey in complete darkness by determining its distance and movement direction. As show in Figure 4.5, bats produce loud pulses and other bats will hear the echoes that bounce back after the pulses hit the prey and surrounding objects. The pulses vary based on species and obstacles encountered.

According to studies, the microbat species builds a virtual three-dimensional scenario of its surroundings based on the loudness variations of the echoes and the time delays between emissions and receptions of echoes. Bats use the Doppler effect of an echoed signal to detect the type of prey, its distance, orientation, and moving speed. Echolocation has inspired many researchers to design and develop novel algorithms for non-telecommunication applications [2] [31] [34]. This ability of bats will be used in the design and development of a novel hybrid routing algorithm referred to as bat-termite for telecommunication domain applications in general and in particular for MANETs.

Echolocation in Bat-Termite
In MA-termite, when the pheromone of a neighbor node which is still within communication range reaches its minimum value (PH]LOOR) in the TPT, it cannot be used as an alternate route to reach the destination. The results are poor network exploration and improper backup route maintenance.

Under this scenario, before removing the neighbor node's entry in the TPT, the battermite algorithm sends a special echolocation packet toward the neighbor node (Figure 4.6a) in unicast mode to observe its current status. Upon receiving the echolocation packet, the neighbor node "piggybacks" its

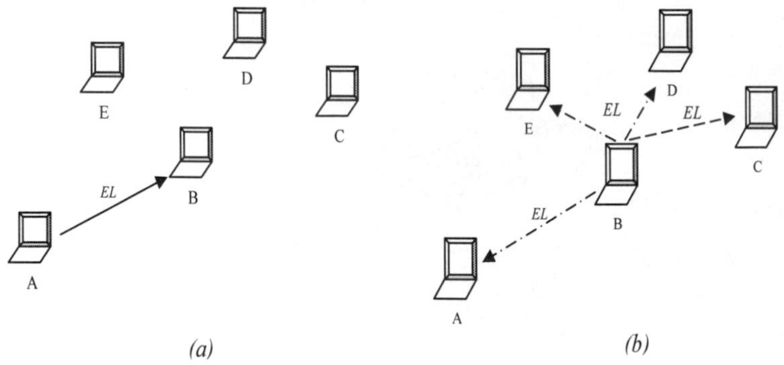

FIGURE 4.6
Echolocation in Bat-Termite

current status (available bandwidth, queue status, distance) in the echolocation packet and broadcasts the packet to its neighbors (Figure 4.6b).

Like bats that collect movement, direction, and distance information about their prey from echoed signals, the bat-termite algorithm discovers the current bandwidth, queue and distance data of the neighbor node using a broadcast echolocation packet. If the neighbor node reports its current status, it will be considered for entry in the TPT. In this way, the echolocation capability of bat-termite maintains fresh entries in the TPT via network exploration and reflects current status of the network.

Bats, during the echolocation process, leak the information to other species as well and this leakage of information may be intentional or unintentional [2]. Study shows that the echoed signals will be overheard by other bats and animals for identifying the surrounding objects. This feature of bats is also exploited in the bat-termite algorithm by means of broadcasting echolocation packets by neighbor nodes as shown in Figure 5.4b. The other nodes upon receiving the broadcasted echolocation packet will also receive the neighbor status and reconsider the broadcasting node entries in the TPT.

Echolocation Characteristics of Bat-Termite

During the echolocation phase of bat-termite, whenever a mobile node receives a broadcasted echolocation packet from its neighbor node, it extracts the following:

1. The neighbor node's distance using the cross layer model (Subsection 4.4.1)
2. The neighbor node's orientation such as whether it is moving toward or away from the node or is stable

3. The mobility information of the neighbor node for determining whether the neighbor node is reliable or not
4. The current status of the neighbor node in terms of available bandwidth, queue status, energy etc. is also extracted using the piggybacked information from the broadcasted echolocation packet.

Route Maintenance Phase of Bat-Termite

Unlike basic termite and MA-termite routing algorithms, the proposed hybrid bat-termite algorithm has a new route maintenance mechanism for issuing a one-hop route error packet (RERR), thereby notifying the path loss to the immediate neighbor. If a node is left with an empty neighbor list while forwarding the data packet toward the destination, a will send a one-hop RERR packet to its immediate neighbor node from which the data packet arrived. This one-hop RERR packet causes the recipient node to remove the RERR source node entry from its TPT and thus drop the RERR packet. The recipient node then finds an alternative route from its remaining neighbor list and forwards the stored data packet toward the destination. If the receipient node cannot find the alternate route to the destination, it will send a one-hop RERR packet to its previous node in the data path. If the recipient node is the source node, it will re-initiate the route setup phase by issuing RREQ packets. The route maintenance and network exploration features of bat-termite are explained in Algorithm 4.3. The structure of the RERR and echo location packets will be discussed in the next section.

Packet Structure of Bat-Termite

The bat-termite algorithm utilizes two additional control features: RERR and echolocation packets. The echolocation packets are event based and similar to the HELLO packets of the traditional AODV protocol. Additional echolocation packets are used to determine the current status of neighbor nodes. The main purpose of the echolocation packets is broadcasting current status (energy, available bandwidth, queue status) of a node to its neighbor nodes.

Echolocation packets in bat-termite do not include status information about their neighbors. Their main objective is informing neighbors of their existence. An echolocation packet mimics the working principle structure of the HELLO packet in the AODV protocol. The RERR packets ofthe bat-termite algorithm mirror the RERR packets of traditional AODV. The only modification is that RERRs in bat-termite only have one-hop lives.

Algorithm 4.3 explains the route maintenance and network exploration aspects of the bat-termite algorithm.

Algorithm 4.3 : Route Maintenance and Network Exploration Aspects of Bat-Termite Algorithm

Input : TPT and LST
Output: Updated TPT and LST for each neighbor

1. ***When*** $p^{x}_{A,d}$ reaches *PH_FLOOR*
2. send *Unicast_echo_location* packet to node A from node x

1. ***When*** node A receives *Unicast_echo_location* packet
2. Update its current status in *Unicast_echo_location* packet
3. Broadcast updated *Unicast_echo_location* packet to all its neighbors

1. ***If*** *Broadcast_echo_location* packet is received at x from A
2. call ***estimate*** = ***Extract_Info(Broadcast_echo_location)*** function
3. Based on value of ***estimate***, update $p^{x}_{A,d}$ in x's TPT
4. ***Else***
5. Remove $p^{x}_{A,d}$ in x's TPT
6. ***Endif***

1. ***If*** *Broadcast_echo_location* packet is received by neighbor n of A
2. call ***estimate*** = ***Extract_Info(Broadcast_echo_location)*** function
3. Based on value of ***estimate***, update $p^{n}_{A,d}$ in n's TPT
4. ***Endif***

1. Extract_Info(broadcast_echo_location)
2. Process broadcast_echo_location packet
3. Extract ***Bandwidth, Energy , Queue status*** information
4. Save extracted information in variable *estimate*
5. Return *estimate*

The bat-termite algorithm was designed to use the echolocation capabilities of bats to overcome the drawbacks of the MA-termite algorithm (poor network exploration and backup route maintenance). The echolocation feature of the bat-termite algorithm keeps all the active node entries that lead to the destination within the TPT. Quick route discovery, high robustness level, efficient management of multiple routes, and rapid route repairs are the main advantages of the bat-termite algorithm. Its reliable path and effective backup route maintenance capability allow the bat-termite algorithm to maintain steady data transmission between sources and destinations, thus producing fewer end-to-end delays. Bat-termite reduces control packet overhead through its robustness, and fewer link breakups due to stable mobile nodes along its paths.

The bat-termite algorithm has been simulated in NS-2 for quantitative and qualitative analyses and the scenarios used to evaluate the MA-termite algorithm were used. Bat-termite performance results were compared to those ofbio-inspired (SARA and basic termite) and non-bio-inspired (AODV) routing protocols. QoS metrics such as throughput, packet drops, end-to-end-delays, and control packet overhead were also compared. The simulation results indicated that bat-termite outperformed the SARA, basic termite, and AODV algorithms when node density was low.

However, the simulation results also indicated that although bat-termite solves the poor network exploration and improper route maintenance problems of MA-termite, neither bat-termite nor MA-termite addressed the stagnation problem. Bat-termite produces additional echolocation control packets under high node density conditions and this deteriorates its performance.

To overcome these drawbacks, a novel load-balanced hybrid bio-inspired routing protocol for MANETs based on the activities of termites and bats has been proposed. It is known as the load-balanced bat-termite (LB-bat-termite) algorithm and is intended to mitigate the stagnation problem of the bat-termite algorithm.

4.4.3 Load-Balanced Bat-Termite: Novel Hybrid Bio-Inspired Routing

This section discusses the state of the art of a heuristic hybrid bio-inspired routing protocol for MANETs. The protocol utilizes a load balancing approach and is known as the load balanced bat-termite (LB-bat-termite) algorithm. It is an enhanced version ofthe MAtermite and bat-termite algorithms in that it considers nodal load to achieve load to achieve load balancing among the available paths. This mitigates the stagnation problem suffered by MA-termite and bat-termite. LB-bat-termite also overcomes the poor backup and route maintenance problems of MA-termite by efficiently managing the multiple paths in the TPT. The details of LB-bat-termite are discussed in the following sections.

Motivation for Development of LB-Bat-Termite

To find the optimal path between the source and destination nodes, the existing protocols for MANETs use many routing metrics such as fewer end-to-end delays and maximum bandwidth in general and the route hop count in particular. The route hop count is the most popular metric for finding the shortest path to a destination node [10].

During a data session, the nodes that appear on the shortest path will get maximum loads compared to loads on other mobile nodes. The maximum loads consume most of their bandwidth, energy, and memory resources. Eventually the heady loads create a bottleneck along the shortest path. This leads to congestion and eventually the network reaches the convergence or equilibrium state. Congestion causes packet loss; in the worst case, heavy packet losses will result in reduced throughput and may lead to connection failure.

The major drawback of any ACO- or termite-based bio-inspired routing algorithm is stagnation which occurs when a network reaches its convergence or equilibrium state. MA-termite and bat-termite algorithms suffer from the

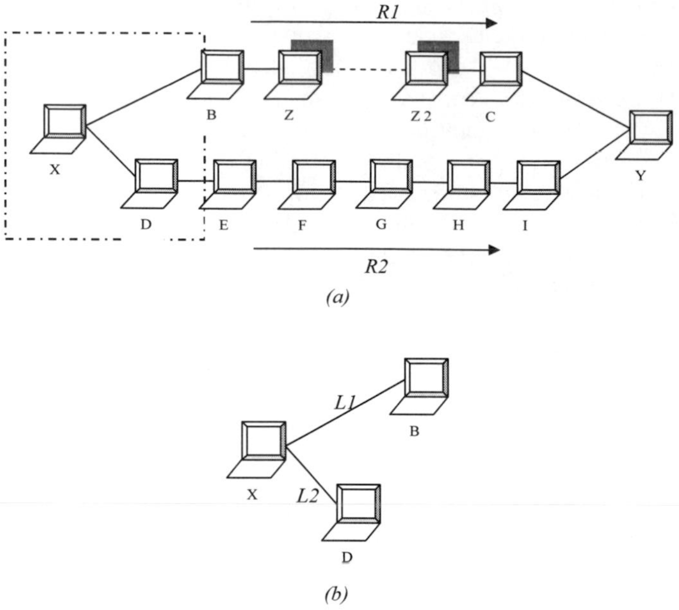

FIGURE 4.7
Two-Path System for Studying Asymptotic Behavior of MA and Bat-Termite

stagnation problem. To illustrate their asymptotic behavior, consider the two-path system shown in Figure 4.7a. Source node X wants to communicate with destination node Y. To reach Y, the source node X has two independent paths designated $R1$ and $R2$. Each path consists of a series of relay nodes with different mobility, bandwidth, and other characteristics. Figure 4.7b shows the subset of Figure 4.7 a. Node X can choose link $L1$ or $L2$ to reach node Y.

Assume during route setup that node X chooses link $L2$ to reach the destination node Y. The trail pheromone on link $L2$ increases. Thus the link is reinforced and it attracts all the traffic. Conversely, $L1$ is negatively reinforced and its pheromone gradually decays over time. Under these circumstances, the network appears to have a single dominant link ($L2$ in this example) between the source and destination nodes.

The dominant links in the MA-termite and bat-termite algorithms will always be chosen. As soon as one link dominates the others, all traffic is diverted toward the dominant link and a bottleneck is created. The less dominant $L1$ link will lose its priority; over time it may be congested by packets and become non-optimal.

Packets are attracted to the link with the most pheromone. In the example above, the trail pheromone on link $L2$ is at a high level. The packets will still choose $L2$ to reach their destination despite the congestion. The congestion in $L2$ will lead to heavy packet loss and low throughput of the system; this condition is known as stagnation and as a result:

1. The probability of selecting the next optimal link decreases (L1 in the above example).
2. The optimal link becomes non-optimal due to congestion (L2 in the above example).
3. The non-optimal link may become optimal (L1 in the above example).

Since mobile nodes of MANETs are resource constrained devices, a good load balancing strategy which uses the available resources of mobile nodes effectively and efficiently is required. In telecommunication networks, the load is broadly classified as, channel load (traffic in the channel), nodal load (packets waiting in the queue, processing of incoming packets) and load generated by the neighbor nodes such as HELLO packets. Nodal load is a node's activity indicating how busy a node is. The nodal load is described in terms of processing, power, memory, and bandwidth loads. In general the existing load balancing protocols can be broadly classified as delay, traffic- or hybrid-based. The load balancing protocols consider metrics such as active paths, traffic size, packets in the interface queue, channel access probability and node delay to achieve load balancing in MANETs [6].

Even though most of the existing routing protocols for MANETs address the load balancing issue, they target load balancing only during the route setup phase. Further, these context unaware routing protocols will not adequately deal with issues such as scalability and dynamic topology. Hence, the current trends have focused on SI for providing load balancing in MANETs. Even though SI is a promising technology and could be used to optimize the network parameters, it suffers from stagnation caused by always choosing the highest pheromone link at any given time. The stagnation arises when the optimal link selected for data transfer is congested over time and reduces the throughput. Thus, stagnation is unacceptable and can be avoided using different techniques namely aging, pheromone smoothing and limiting, evaporation, privileged pheromone laying, and pheromone heuristic control [29]. These approaches reduce the influence of past experience and allow new and better links for effective data transfer. LB-bat-termite addresses the stagnation problem by balancing the load among the available multiple paths using pheromone heuristic control. The details of load balancing are discussed below.

Load Balancing Approach of LB-Bat-Termite

The proposed LB-bat-termite introduces another kind of pheromone called alarm pheromone which is used by the termites to indicate danger. Alarm pheromone in LB-bat-termite indicates the load on each neighbor node and is defined as a function of total number of packets waiting in the queue [12]. LB-bat-termite defines a new forwarding function based on heuristics for finding

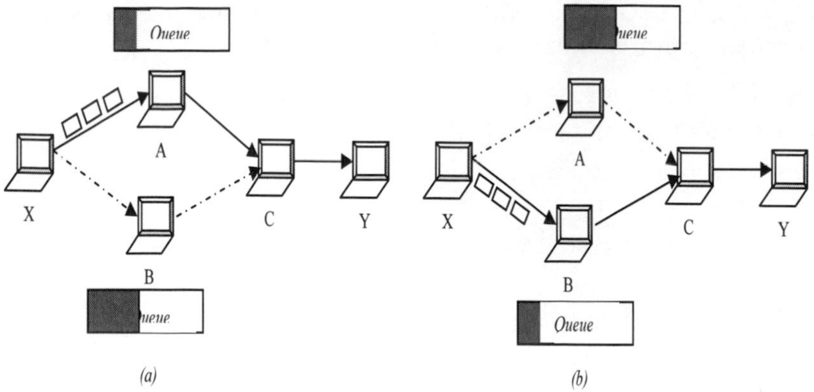

FIGURE 4.8
Load Balancing Approach of LB Bat-Termite

the probability of neighbor nodes and forwards packets toward the corresponding destination node. LB-bat-termite also defines a new heuristic function for updating the alarm pheromone on each neighbor link to reflect the current load on the neighbor node.

The proposed LB-bat-termite algorithm avoids the links with higher alarm pheromone (link with high load) concentrations and chooses alternate links to reach the destination before it gets congested. Thus, LB-bat-termite avoids congestion by neglecting the congested link during the data session phase. During the route discovery phase, LB-bat-termite considers only the trail pheromone for finding the multiple paths toward the destination while a combination of alarm and trail pheromones is used during the data session phase for forwarding the data packets. Thus the proposed LB-bat-termite algorithm achieves load balancing during both route discovery and route maintenance phases, thereby providing high robustness. Further, LB-bat-termite does not need additional control packets for finding the neighbor node's load as queue status is piggybacked in all outgoing packets. Figure 4.8 depicts the load balancing approach of the proposed algorithm.

As shown in Figure 4.8a, let us consider that a node X initially chooses a neighbor node A with less load for reaching the destination. When node$A$ has a higher load than node B, node X switches the traffic to node B as shown in Figure 4.8b, and LB-bat-termite achieves dynamic load balancing. The alarm pheromone update function is explained in the following section.

Alarm Pheromone Update Function of LB-Bat-Termite

LB-Bat-Termite uses both control and data packets for updating alarm pheromone in a given node. Whenever a mobile node receives a packet from the neighbor node, it first extracts the piggybacked queue status from the packet and then calculates the alarm pheromone for the corresponding neighbor node using 6.1.

$$\phi_i^n = 1 - \frac{Total\ No.\ of\ Packets\ in\ Queue}{Q\_MAX} \tag{4.3}$$

where ϕ_i^n is the amount of alarm pheromone of neighbor node i at node n and Q_MAX is the maximum number of packets that the queue can accommodate. The calculated alarm pheromone will be stored in a table on the routing layer for further processing. To check the freshness of the alarm pheromone entry in the table, the time t at which alarm pheromone is calculated will also be stored in the same table on the routing layer. The differences between the alarm pheromone and trail pheromone are explaced in the next section.

Difference between Trail Pheromone and Alarm Pheromone
The comparative features of trail and alarm pheromones are given below:

1. Alarm pheromone over the neighbor links will not decay over time.
2. For finding all available paths between the source and the destination, trail pheromone will be used. For finding the reliable and less congested path among the available paths to the destination node, a combination of trail and alarm pheromones will be used.
3. Data packets will be used for updating trail pheromones over the link whereas both data and control packets will be used for updating alarm pheromones over the link.
4. The trial pheromone is a function of total number of packets passing through the node, and an alarm pheromone is a function of total number of packets in the queue.

The computed alarm pheromone will be further used by the forwarding function to estimate the next hop probability as discussed below.

Forwarding Function of LB-Bat-Termite The proposed heuristic model of LB-bat-termite uses the the trail pheromone $P_{i,d}^n$ and a heuristic function of alarm pheromone η_i^n to calculate the next hop probability. At each mobile node, the packet can be forwarded as a functional composition of $P_{i,d}^n$ and η_i^n. The heuristic function is:

$$\eta_i^n = \frac{\phi_i^n}{\sum_{k \in N} \phi_k^n} \tag{4.4}$$

where ϕ_i^n is the alarm pheromone for the neighbor node i at node n and N is the total number of neighbor nodes of node n. The forwarding function is defined by:

$$p_{i,d}^n = \frac{P_{i,d}^n (\eta_i^n)^F}{\sum_{k \in N} P_{k,d}^n (\eta_k^n)^F} \tag{4.5}$$

where $P_{i,d}^n$ is the trail pheromone of node i to reach the destination d at node

n; N is the total number of neighbor nodes of node n; η_i^n is the heuristic value calculated based on an alarm pheromone for neighbor node i at node n. The constant F is defined as the inertia factor and it can be used for controlling the routing behavior of LB-bat-termite. The preferences for the next hop selection can be varied by choosing the proper value of F; a lower value of F can be considered generally for selecting the highest trail pheromone link for forwarding the data packets. On the other hand, a higher value of F orients the data packets toward more optimistic heuristic values, but the optimal value of F is dependent on the pheromone ceiling *PH_CEILING*. The node structure of LB-bat-termite is explained below.

4.4.4 Node Structure of Load-Balanced (LB) Bat-Termite

The LB-bat-termite, MA-termite, and bat-termite algorithms maintain two tables on the network layers in each node: the local status table (LST) and trail pheromone table (TPT); see Section 4.2.4).

Local Status Table

While forwarding the data packets toward the destination node, each node will refer to its LST to extract current status of neighbor nodes in terms of alarm pheromone and current and previous distances. Based on the freshness of the data (time interval), each node decides whether to consider the entry in the LST for forwarding the data. To remove the stale entries in the LST, the contents are refreshed every *LST_TIME_OUT* seconds. The current and previous distances of neighbor nodes will give the mobility information that is computed as follows.

Neighborhood Node Distances: Whenever a node receives a packet, it first calculates the distance of the transmitting node using the free space propagation model (Subsection 4.2.1). The calculated distance d along with the time interval t at which the distance is calculated is stored in the LST of the current node. Thus, every node will be aware of its neighbor node mobility, movement direction, and chances of staying within the transmission range. Alarm pheromone can be calculated to reflect the load on each neighbor node and the details of computing alarm pheromone are given below.

Alarm Pheromone: Whenever a node receives a data or control packet, it first calculates the alarm pheromone based on the total number of packets in the sending node's queue using Equation (4.3) and this alarm pheromone along with the time interval t at which the alarm pheromone is calculated are stored in the LST. The stored information of LST will be used by the forwarding function to estimate the next hop probability. The trail pheromone update and decay function will also use the LST entries to update the trail pheromone in the TPT of each node. The Algorithm 4.4 gives the alarm pheromone update function of LB-termite and the TPT structure is shown in the next section.

Algorithm 4.4: Alarm Pheromone Update Function of LB-Bat-Termite

Input: LST
Output: An updated LST with current network status
1. At ***node x*** in network
2. ***For every packet*** received by x from neighbor node
a. Extract queue information piggybacked in packet
b. Calculate alarm pheromone for corresponding neighbor node
c. Calculate current distance of neighbor node
d. Update x's LST with calculated information for neighbor node
e. Piggyback queue status of x in packet
3. Forward packet

Trail Pheromone Table

The trail pheromone table (TPT) is used for maintaining the trail pheromone for each reachable destination node. Each node records the amount of destination trail pheromone on each neighbor link in the TPT and the process is analogous to routing tables of traditional routing protocols such as AODV. The TPT contains the same elements as traditional routing tables: neighbor node identification, destination node identification, and routing metric (pheromone in LB-termite and number of hops in AODV). The rows represent neighbor nodes whereas the columns correspond to the destination nodes.

When a node detects a new neighbor, it adds an additional row in the TPT; whenever a new destination is discovered, a new column is added in the TPT and in both the cases the trial pheromone will be initialized to *PH_INITIAL*. While updating the trail pheromone of a neighbor node on the TPT, each node has to refer to the corresponding entries in the LST for finding the alarm pheromone and current and previous distances of the neighbor node to reflect the mobility and load information of the neighbor node.

Since each node movement is reflected in the TPT, the neighbor node with low mobility will have a higher trail pheromone entry and the neighbor node with high mobility will have less trail pheromone entry. Further, load on each neighbor will also be reflected on the TPT based on the alarm pheromones of each neighbor node. Algorithm 4.5 gives the trail pheromone update function of LB-bat-termite. Algorithms 4.6, 4.7, and 4.8 give the route setup phase, route maintenance phase and data session phase of LB-bat-termite. The TPT and LST of each node can be updated during the route discovery and maintenance phases as explained in the following subsections.

Load-balanced bat-termite (LB-Termite) finds the stable and reliable nodes for the path and tries to mitigate stagnation by introducing a new kind of pheromone called the alarm pheromone. High route reliability, quick route discovery, prompt repair and maintenance, high robustness with efficient management of multiple routes and stagnation avoidance are some of the features of the algorithm. LB-bat-termite is implemented in NS-2 and the results are compared with other state-of-the-art bio-inspired (SARA and termite

Algorithm 4.5 Trail Pheromone Update Function of LB-Bat-Termite

Input: TPT

Output: Trail pheromone for source and destination nodes will be updated in TPT

For every data and RREQ packet received at node x from neighbor node y

1. ***If*** data packet
2. ***for*** all neighbors
3. Decay pheromone
4. ***end for***
5. ***End if***
6. Extract distance and ϕ_x^y information from LST for node y
7. Calculate $P_{y,d}^n$ for corresponding neighbor
8. Update $P_{y,d}^n$ for y in x's TPT
9. Piggyback queue status of current node in packet

algorithms) and non-bio-inspired (AODV) routing algorithms for performance evaluation. The same simulation environment considered for evaluating LB-bat-termite is considered for evaluating MA-termite and bat-termite. LB-bat-termite achieves a 2% to 4% increase in throughput, end-to-end delay and total packet drops when compared to SARA and achieves more than 4% increase in throughput, end-to-end delay, total packet drops and control packet overhead when compared to termite and AODV algorithms when node density is low (up to 200).

As node density increases (above 200), LB-bat-termite suffers from additional control packet overhead, thereby deteriorating the performance. Further, the resultant graphs clearly depict that LB-bat-termite under high node density shows poor performance when compared to SARA and termite algorithms. Thus, even though LB-bat-termite algorithm solves the stagnation problems of MA-termite and bat-termite and solves the poor network exploration and backup route maintenance problems of MA-termite, it suffers from additional control packet overhead in maintaining the active paths toward the destination nodes. To overcome the control packet overhead problem, a novel heuristic load-balanced bio-inspired routing algorithm known as load-balanced termite will be discussed in the next section.

4.5 Load-Balanced Termite: Novel Load Aware Bio-Inspired Routing

This section discusses the working principles of a novel heuristic bio-inspired protocol for MANETs that incorporates load balancing. It is known as the load-balanced termite (LB-termite) algorithm [18] and considers nodal loads

Algorithm 4.6: Route Setup Phase of LB-Bat-Termite

Input: TPT and LST

Output: Updated TPT and LST and optimal path between source and destination nodes

1. ***If*** node n does not find p_d^n in its TPT to forward data packets
2. Buffer data packets
3. Release *RREQ* packets from node n to find new pheromone trail to destination

1. ***When*** an intermediate node receives *RREQ* packet
2. ***if*** it is destination or consists of pheromone trail to destination
3. Unicast *RREP* packet to the source
4. Discard *RREQ*
5. ***else***
6. Update alarm and trail pheromone in TPT and LST
7. ***if*** neighbor exists
8. Piggyback queue status information in packet
9. Randomly choose neighbor and forward RREQ packet
10. ***else***
11. Discard RREQ
12. ***End if***
13. ***End if***
14. ***End When***

1. ***If*** node n receives *RREP* packet within *RREQ_TIME_OUT*
2. Forward buffered data packets and start data session
3. ***Else***
4. Send *RREQ* packets from node n to find new pheromone trail to destination
5. ***End if***

to achieve load balancing among the available paths, thereby mitigating the stagnation problems of the MA-termite and bat-termite algorithms. The following sections discuss the details of LB-termite.

4.5.1 Motivation for Development of LB-Termite Algorithm

The LB-bat-termite algorithm discussed earlier suffers from additional control packet overhead caused by its echolocation packets. The algorithm uses the packets to overcome the poor network exploration and improper backup maintenance problems of the MA-termite algorithm.

LB-bat-termite uses echolocation packets to maintain all possible paths to the destination node in its TPT. Load balancing across multiple paths ensures that LB-bat-termite will not allow trail pheromone levels of primary

Algorithm 4.7: Route Repair Phase of LB-Bat-Termite

Input: LST and TPT

Output: 1. Neighbor node's ϕ and its distance updated in LST
2. P for neighbor node removed from TPT

1. ***For*** every RERR packet received at n from neighbor node y
2. Update ϕ_x^y in LST
3. Remove $P_{y,d}^n$ from *n's* TPT
4. ***If*** neighbor node for y leads to destination
5. Calculate next-hop probability for all neighbors which lead to destination
6. Choose highest probability neighbor and forward data packet
7. Drop RERR packet
8. ***Else***
9. Drop RERR packet
10. Send new one-hop *RERR* packet to next node toward source
11. ***End if***
12. ***End for***

1. ***If*** n is source and receives RERR packet
2. Update ϕ_x^y in LST
3. Remove $P_{y,d}^n$ from *n's* TPT
4. Buffer data packet
5. Trigger route setup phase
6. Drop RERR packet
7. ***End if***

and secondary paths to fall below PH_FLOOR in the TPT. Thus, only the primary and secondary paths among all the possible paths to the destination will be used for data transfer at any point in time. To illustrate this concept. consider the n path system shown in Figure 4.9a in which source node X wants to communicate with destination node Y.

To reach Y, the source node should have n independent paths designated $Rl, R2, ...Rn$. Each path consists of a series of relay nodes connecting source node X to destination node Y. The relay nodes have different bandwidths, mobilities, and other characteristics. Figure 4.9b shows the subset in which the source node X has many links $(Ll, L2, L3, ...Ln)$ by which to reach destination node Y.

Suppose node X chooses $L2$ as its primary link to the destination during route set up and let $L3$ be secondary among the available links to destination node Y. The load balancing ability of LB-bat-termite efficiently manages the load across the primary and secondary links so that the pheromones on these links will never fall below PH_FLOOR in the TPT of node X The pheromones on the other links $(Ll, L4, ...Ln)$ will decay gradually over time and reach PH_FLOOR.

Algorithm 4.8: Data Session Phase of LB-Bat-Termite

Input: TPT and LST

Output: Updated TPS and LST and reinforcement of optimal path between source and destination

```
while data packet reaches destination node
    At node x which is intermediate node in path
        for all neighbors of node x
            Decay pheromone
        end for
        for neighbor y from which data packet arrived
            Update p^x_{s,d} in x's TPT where s is source node
            Update φ^y_x and distance of y in x's LST
        Update path utility value in data packet
        Calculate next-hop probability for all neighbors of x which lead to destination
        Choose highest probability neighbor and forward data packet
        if no neighbor exists for x which lead to destination
            Buffer data packet
            Initiate route setup phase
        end if
end while
```

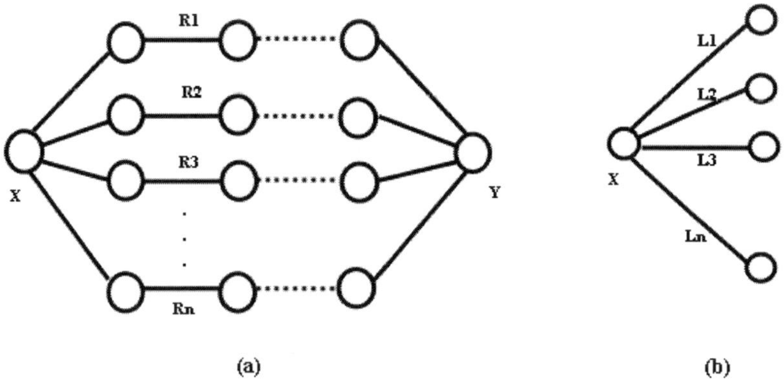

FIGURE 4.9
Paths and Links of LB Bat-Termite

Under these circumstances, the source node X generates echolocation packets on links $Ll, L4, ...Ln$ to determine the current status of its neighbor nodes and thus maintains the backup route to destination node Y. Although these backup links are maintained and refreshed regularly in the TPT, the chances of using these links for data transfer decrease because the primary and secondary links will dominate all the others. However, regular refreshing of the backup links generates additional echolocation packets, thus increasing the network load.

The nodes in LB-bat-termite try to maintain all possible routes to the destination and thus spend most of their time refreshing TPT instead of data transfer. These additional echolocation packets increase the load on the network and decrease performance. LB-termite will address this additional control packet overhead issue and the details are discussed in the following sections.

4.5.2 Working Principle of LB Termite Algorithm

To reduce the additional control packet overhead problem of the LB bat-termite algorithm, the proposed LB termite algorithm drops the echolocation feature while retaining the remaining LB bat-termite features. LB termite uses Equation (4.1) to calculate trail pheromone update and decay and uses Equation (4.3) to update the alarm pheromone over the links. LB-termite uses Equations (___) and (___) as forwarding functions. It uses Algorithms ___ and ___ for updating alarm and trail pheromones in the TPT and Algorithms ___' ___' and ___ for route discovery and maintenance and data sessions.

4.6 Comparative Study of Mobility Aware (MA) Termite, Bat-Termite, LB Bat-Termite and LB Termite Algorithms

The MA-termite (MA-T), bat-termite (bat-T), LB bat-termite (LB-Bat-T) and LB termite (LBT) algorithms are simulated in NS-2. The performance results were compared with bioinspired (SARA [7] and basic termite algorithms (BT) [27] [28] and with the non-bioinspired ad hoc on demand distance vector (ADDV) [5] routing algorithm. The QoS metrics (throughput, packet drops, end-to-end delays, and control packet overhead) were compared. The rigorous study was conducted under two scenarios to determine how the proposed algorithm would perform with regard to scalability and node velocity.

Scenario A The scalability of the proposed algorithm was evaluated by examining two test areas (500 m × 500 m and 1000 m × 1000 m. For each simulation area the algorithm was tested at densities of 100, 200,400, and 600 nodes. Node velocity was kept constant at 2 m/s (normal walking speed of a person) and no pause times were given to mobile nodes.

Scenario B The performance of the proposed algorithm at different node velocities was studied. Mobile nodes of varying velocities (2, 4, 6, 8, and 10 m/s) and densities (100, 200, 400, and 600 nodes) were tested during each simulation run. The test area measured 1000 m × 1000 m and no pause times were given to mobile nodes.

Tables 4.1 and 4.2 summarize the QoS results for the various algorithms under different conditions of scalability and node velocity.

Bio-inspired routing, an attractive and promising technique, is outperforming the traditional routing protocols by maintaining high throughput with less control packet overhead. However, the bio-inspired algorithms such as ACO and termite-based routing suffer from stagnation which leads to congestion in the optimal path and reduces the throughput. The proposed bio-inspired routing algorithm based on termite and load balancing referred to as load-balanced-termite (LB-termite) finds the stable and reliable nodes for the path and tries to mitigate stagnation by introducing a new alarm pheromone. High route reliability, quick route discovery and repair and maintenance, high robustness with efficient management of multiple routes and stagnation avoidance are some of the highlights of the LB-termite algorithm. The algorithm was implemented in NS-2 and the results were compared with other state-of-the-art bio-inspired (SARA and termite algorithms) and non-bio-inspired (AODV) routing algorithms to examining performance evaluation. LB-termite achieved 3% to 5% of increase in metrics throughput, end-to-end delay and total packet drops when compared to SARA and a more than 5% increase in throughput, end-to-end delays, total packet drops and control packet overhead when compared to the termite and AODV algorithms.

The proposed LB-termite algorithm can be applied in a virtual classroom (VCR) scenario. VCR is a teaching and learning process that connects a group of students and a teacher through the Internet. A VCR session can be established immediately and students may join or leave the session as they wish. A teacher can establish a session from his office or home. Students can join sessions from any place on the campus. They can send instant messages to other group members, request study materials, and transfer files.

In such a situation, LB-termite's efficient route maintenance and multi-path features can ensure a constant data rate among the group members. Its ability to reduce control packet overhead means LB-termite improves network lifetime; because it allows fewer packet drops, it avoids session hiccups. Fewer end-to-end delays make VCR more interactive and lively for students and teachers.

The LB-termite algorithm can also be used for other academic application scenarios such as information management and real-time live data sharing during classroom lectures and video conferences.

TABLE 4.1
Protocol Behavior under Increasing Node Density (100 to 600 Nodes)

Algorithm	**Throughput**	**Drops**	**End-to-End Delay**	**Control Packets**
AODV	Low	High	High	High
Termite	Medium	High	High	High
SARA	High	Low	Low	Low
MA-T	High	Medium	Medium	Medium
Bat-T	Low	High	High	High
LB-Bat-T	Low	High	High	High
LB-T	High	Low	Low	Low

TABLE 4.2
Protocol Behavior under Increasing Node Velocity (2 to 10 m/s)

Algorithm	**Throughput**	**Drops**	**End-to-End Delay**	**Control Packets**
AODV	Low	High	High	High
Termite	Medium	High	High	High
SARA	Medium	Medium	Low	Low
MA-T	High	Medium	Medium	Medium
Bat-T	Low	High	High	High
LB-Bat-T	Low	High	High	High
LB-T	High	Low	Low	Low

4.7 ACO-Based QoS-Aware Routing Algorithms

Ant-Based Distributed Routing Algorithm (ADRA) (2004)
Zheng et al. developed a QoS aware ACO-based bio-inspired routing algorithm known as ADRA for MANETs application. The algorithm has some features similar to those of the probabilistic emergent routing algorithm (PERA); the differences lie in checking available resources before forwarding ants and path setup only after the system meets QoS requirements. Along with forward and backward ants, ADRA uses two more types known as enforce ants and anti-ants for fast convergence. The authors demonstrated through simulation that ADRA outperforms non-bio-inspired DSR algorithms.

Ant Routing Algorithm for Mobile Ad Hoc Networks (ARAMA) (2005)
Hussein et al. [13] developed a novel QoS aware bio-inspired routing algorithm referred to as ARAMA which uses multiple performance metrics such as node energy, processing power and link bandwidth. ARAMA manages the resources through fair distribution and continuously checks for new and better paths using forward and backward ants. In addition to forward and backward ants, the algorithm uses negative backward ants to de-emphasize the negative paths and destination trail ants to increase connection setup time. Old solutions are removed through pheromone evaporation. Simulations proved that ARAMA performs better than AODV and is suitable for energy management in MANETs.

Swarm-Based Distance Vector Routing (SDVR) (2007)
Asokan et al. [3] developed a QoS aware bio-inspired algorithm derived from AntNET [9] and known as SDVR. It targets the QoS metrics of delay, jitter, and energy. It maintains a pheromone table for each QoS metric for making routing decisions and thus provides a multi-constrained end-to-end path between the source and destination. SDVR uses pheromone evaporation to remove old paths. Simulation results indicate that it outperforms AODV.

Fuzzy-Based ACO Algorithm (FACO) (2009)
Goswami et al. [23] developed a novel bio-inspired routing algorithm for MANETs referred to as FACO, It combines fuzzy logic with ACO principles. Energy consumption rate, packet buffer occupancy rate, and signal strength are fuzzified to compute the pheromone over a link. FACO uses forward and backward ants for route discovery and setup phases. Each intermediate node calculates the fuzzy cost and updates the forward ants. Whenever the destination node receives a forward ant, then it transfers the stored fuzzy cost to a backward ant and accordingly updates the pheromone table. Simulation results indicate that FACO performs better than ant-AODV [22] and ant colony based routing algorithms (ARA) [30].

Ant Dynamic Source Routing (ADSR) (2008)
Asokan et al. [4] proposed a QoS aware bio-inspired routing algorithm referred to as ADSR and derived from traditional DSR. It targets delay, jitter, and energy. ADSR uses forward and backward ants for route discovery and setup. The forward ant maintains a record of visited nodes and they can be transferred to the backward ant at the destination node. Simulation results have shown that ADSR outperforms DSR.

Multiple Disjoint QoS Enabled Ant Colony-Based Multipath Routing (QAMR) (2012)
Krishna et al [24] proposed a QoS aware multiple disjoint bio-inspired routing algorithm referred to as QAMR that targets the bandwidth problem of MANETs. MQAMR uses bandwidth, delay, and hop count to calculate the path preference probability. Forward ants use the next hop availability (a function of mobility and energy factor) during route discovery and track the local information of visited nodes. Whenever the destination node receives the forward ant, it calculates the path preference value based on the collected information and accordingly the backward ant that satisfies the QoS requirements will be generated for the path. Then the backward ant updates the pheromone value of each visited node proportional to the computed path preference value. The authors demonstrated that QAMR performs better than AODV and ARMAN under high traffic loads.

QoS Aware Ant-Based Multipath Routing Algorithm (2012)
Kim [16] proposed an ACO-based QoS aware routing protocol for MANETs based on the load balancing strategy for maintaining the QoS requirements of the path. Each ant chooses the next hop based on the communication adaptability (function of distance between nodes and queue length) and pheromone will be updated in each visited node. Among the available paths, the most adaptable one is selected based on the bandwidth and delay requirements and the remaining available paths are considered for load balancing. Routing adaptability is calculated based on path energy levels for deciding the data distribution among paths. The proposed algorithm proved better than ant colony multipath routing (AMPR) and colony-based multipath routing (CMPR).

4.8 Summary

This chapter explains the states of the art of various SI-based routing algorithms that provide QoS for MANETs end users. It concentrates on termite- and ant-based QoS aware routing algorithms and highlights hybrid approaches for providing QoS.

The chapter starts by explaining termite-based algorithms that employ CLD approaches to provide QoS. It discusses the MA-termite algorithm that finds a stable resource-rich path between the source and destination nodes. It also covers the bat-termite algorithm that overcomes the improper backup route maintenance problem of the MA-termite algorithm. Although MA-termite and bat-termite outperform traditional and other bio-inspired algorithms, they fail to solve the stagnation and improper network exploration problems. LB-bat-termite was designed to overcome these problems. It works well in low density networks but its performance degrades as node density increases.

Finally, the termite-based routing protocols section discusses LB-termite. This algorithm outperforms MA-termite, bat-termite, and LB-bat-termite in all QoS parameters (throughput, packet drops, end-to-end delays, and control packet overhead). It also solves the stagnation and poor network exploration problems of MA-termite and bat-termite and works well even under increasing node density. The chapter also covers state-of-the-art ACO-based QoS aware routing algorithms for MANETs.

Exercises

1. Write an algorithm to show the route discovery process and route maintenance process of the MA-termite algorithm.
2. List and explain two main drawbacks of the MA-termite algorithm.
3. What is a hybrid algorithm with respect to SI-based routing algorithms.
4. List insects and animals other than bats and termites whose behaviors may be combined to produce a novel routing algorithm for MANETs.
5. Explain why the bat-termite algorithm fails to cope with the increasing node density network.
6. Write a short note about load balancing.
7. Explain the state-of-the-art working principle of the LB-termite algorithm.

References

[1] A.G. Zakeerhusen et al. Cognitive network layer- mobility aware routing protocol. In *Sixth International Conference on Information Processing*, volume 292, pages 84–92, 2012.

[2] Matti Airas. Echolocation in bats. In *Proceedings of Spatial Sound Perception and Reproduction*, 2003.

[3] A Natarajan, R Asokan, and A Nivetha. A swarm based distance vector routing to support multiple quality of service metrics in mobile ad hoc networks. *Journal of Computer Science*, 3(9):700–707, 2007.

[4] A Natarajan, R Asokan, and C Venkatesh. Ant-based dynamic source routing protocol to support multiple quality of service (QoS) metrics in mobile ad hoc networks. *International Journal of Computer Science and Security*, 2(3), 2008.

[5] C.E. Perkins and E.M. Royer. Ad hoc on demand distance vector (AODV) routing. In *Second IEEE Mobile Computing Systems and Applications*, pages 90–100, 1999.

[6] Anh-Ngoc, Chai Keong Toh, and You-Ze cho. Load-balanced routing protocols for Ad Hoc mobile wireless networks. *Topics in Ad Hoc and Sensor Networks*, 47(8):78–84, 2009.

[7] Fernando Correia and Teresa Vazao. Simple ant routing algorithm. In *International Conference on Information Networking, IEEE*, pages 1–8, 2008.

[8] E. Crawley et al. A Framework for QoS-Based Routing in the Internet. Technical report, 1998.

[9] G. Di Caro and M. Dorigo. Mobile agents for adaptive routing. In *System Sciences, Proceedings of 31st International Conference*, volume 7, pages 74–83, 1998.

[10] John Bicket Douglas, S.J. Deel Couto, Daniel Aguayo and Robert Morris. A high-throughput path metric for multi-hop wireless routing. In *Proceedings of 9th ACM International Conference on Mobile Computing and Networking*, 2003.

[11] F. Xia QoS Challenges and opportunities in wireless sensor actuator networks. *Sensors*, 8:1199–1206, 2008.

[12] Praveenkumar Hoolimath et al. Optimized termite: A bio inspired routing algorithm for MANET. In *IEEE International Conference on Signal Processing and Communications*, pages 1–5, 2012.

[13] T. Saadawi, O Hussein, and M. Jong Lee. Probability routing algorithm for mobile ad hoc networks resources management. *IEEE Journal of Selected Areas in Communications*, 23(12):2248–2259, 2005.

[14] Goran Martinovic Josip Balen, Drago Zagar. Quality of service in wireless sensor networks: a survey and related patents. *Sensors*, 8(2):1099–1110, 2008.

[15] KI-IL Kim and Sang-Ha Kim. Effectiveness of reliable routing protocols in mobile ad hoc networks. *Wireless Personal Communications*, 38:377–390, 2006.

[16] Sungwook Kim. An anti-based multipath routing algorithm for QoS aware mobile ad hoc networks. *Wireless Personal Communications*, 66:739–749, 2012.

[17] M Kiran and G Ram Mohana Reddy. Throughput enhancement of the prioritized flow in self aware MANET based on neighborhood node distances. In *IEEE International Conference on Computer Applications and Industrial Electronics*, pages 497–502, 2011.

[18] M Kiran and G Ram Mohana Reddy. Design and evaluation of load-balanced-termite: a novel load aware bio inspired routing protocol for mobile ad hoc networks. *Journal of Wireless Personal Communications,* DOI 10.1007/s11277-013-1453-9, 2013.

[19] M Kiran and G Ram Mohana Reddy. Mobility aware termite: a novel bio inspired routing protocol for mobile ad hoc networks. *IET Journal of Networks*, DOI: 10.1049/iet-net.2012.0203, 2013.

[20] M Kiran and G Ram Mohana Reddy. Bat-termite: a novel hybrid bio-inspired routing protocol for mobile ad hoc networks. *International Journal of Wireless and Mobile Computing*, DOI:10.1504/IJWMC. 2014.062032, 2014.

[21] M. Kiran. Quality of Service Aware Routing Algorithms for Mobile Ad Hoc Networks. PhD thesis, National Institute of Technology Karnataka, Surathkal, India, 2013.

[22] C.K Tham, S Marwaha, and D Srinivasan. Mobile agents based routing protocol for mobile ad hoc networks. In *Proceeding of IEEE Global Telecommunications Conference*, IEEE Computer Society, Washington, pages 163–167, 2002.

[23] R.V. Dharaskar, M.M. Goswami, and V.M. Thakare. Fuzzy ant colony based routing protocol for mobile ad hoc networks. In *Proceedings of International Conference on Computer Engineering and Technology, IEEE Computer Society*, volume 2, pages 438–444, 2009.

[24] V Saritha, P. Venkata Krishna, G Vedha, A Bhiwal, and A.S. Chawla. Quality-of-service-enabled ant colony-based multipath routing for mobile ad hoc networks. *IET Communications*, 6(1):76–83, 2012.

[25] Y Bai. Q Han, L.Gong, and W. Wu. Link availability prediction-based reliable routing for mobile ad hoc networks. *IET Communications*, 5(16):2291–2300, 2011.

[26] Martin Roth and Stephen Wicker. Asymptotic pheromone behavior in swarm intelligent MANETS an analytical analysis of routing behavior. In *Sixth IFIP IEEE International Conference on Mobile and Wireless Communication Networks*, 2004.

[27] Martin Roth and Stephen Wicker. Network routing with filters: link utility estimation in swarm intelligent manets. Technical report, Defense Advance Research Project Agency (DARPA) and Army Research Office under Emergent Surveilliance Plexus MURI Award DAAD 19-01-1-0504, 2004.

[28] N. P. Kulkarni and Samita Indurkar. Congestion control in wireless sensor networks: a survey. *Tazyeen Ahmad International Journal of Engineering Research and Applications,*, 4(11):109–113, 2014.

[29] Kwang Mong Sim and Weng Hong Sun. Ant colony optimization for routing and load balancing: survey and new directions. *IEEE Transactions on Systems, MAN and Cybernetics Part A*, 33(5):560–572, 2003.

[30] P Druschel, D Subramanian and J Chen. Ants and reinforcement learning: a case study in routing in dynamic networks. In *Proceedings of International Joint Conference on Artificial Intelligence*, pages 832–838, 1997.

[31] Leandro dos S. Coelho, Teodoro C. Bora, and Luiz Lebensztajn. Bat-inspired optimization approach for the brushless DC wheel motor problem. *IEEE Transactions on Magnetics*, 48(2):947–950, 2012.

[32] A Kaloxylos, L Merakos, D Vali, S Paskalis. A survey of Internet QoS signaling. *IEEE Communications Surveys & Tutorials*, 6:32–12, 2004.

[33] Wu Mu-Qing, Wu Da Peng, and Zhen Yan. Reliable routing mechanism based on neighbor stability for MANET. *Journal of China Universities of Posts and Telecommunications*, 16(3):33–39, 2009.

[34] Xin-She Yang. A new metaheurastic bat-inspired algorithm. In *Nature Inspired Cooperative Strategies for Optimization, Studies in Computational Intelligence*, Springer, 2010.

[35] W. Guo, X Zheng, and R Liu. An ant-based distributed algorithm for ad- hoc networks. In *Proceedings of International Conference on Communications, Circuits, and Systems, IEEE* Computer Society, Washington, pages 412–417, 2004.

5

SI Solutions to Security Issues in MANETs

CONTENTS

AMANET is an infrastructure-less network consisting of mobile nodes with wireless network interfaces. To enable communications among nodes, MANETs dynamically establish paths among the nodes in a multi-hop manner. MANETs find many applications in modern communications but their nature and structure make them vulnerable to various types of attacks. Attackers use various approaches to acquire the valuable resources of MANETs and the attacks decrease throughput and overall network performance.

MANET attacks can be classified as active and passive. Attackers may be internal or external. An internal attacker is an authorized node within the routing mechanism. An external attacker is not a legitimate part of a network. This chapter overviews various security issues and their solutions. It starts with a list of security vulnerabilities and goals of MANETs,followed by a detailed discussion of the types of attacks on different layers of protocol stacks. Routing and security issues intended to ensure security communications are illustrated and the chapter concludes with an explanation of SI-based security routing.

5.1 Security Goals

MANETs are self-organizing, infrastructure-less, multi-hop networks. Their characteristics such as wireless communication media and distributed nature pose great challenges to system security designers. Nodes in MANETs are self organizing and securing the networks is difficult. The security goals for MANETs are:

1. ***Availability*** ensures survivability of network services regardless of its security state and enables authorized users to access data and services at appropriate times.
2. ***Integrity*** ensures that message sent by the source node reach the destination without content alteration or corruption.
3. ***Confidentiality*** is keeping information secret. The message sent by the sender is kept secret and only the intended receiver authorized to access it can read the message.
4. ***Authentication*** ensures that the participants in communication are genuine and not impersonators.
5. ***Authorization*** assigns access rights to different types of users and specifies privileges and permissions.
6. ***Anonymity*** is closely related to privacy preservation. All the information that can be used to identify the owner or current user of a node should by default be kept private and not be distributed by the node or the system software.
7. ***Non-repudiation*** ensures that sender and receiver of a message cannot disavow sending or receiving such a message. This aids in determining whether a node with some undesired function is compromised.

5.2 Vulnerabilities

Avulnerability is a weakness in a security system. MANETS are more vulnerable than wired networks. AMANET node may be vulnerable to wireless links that are susceptible to attacks aimed to prevent normal communications among nodes. MANET nodes lack defense mechanisms to protect them from internal and external attacks. Furthermore, the variations in processing speed, computational power, and storage capacities of devices such as laptop computers and mobile phones may draw the attention of attackers. Another vulnerability is allowing data access to unauthorized users or failing to verify user identifies. MANETs also exhibit the vulnerabilities listed below.

1. ***Lack of centralized management***: MANETs are large scale ad hoc networks and centralized monitor servers. It is not easy to monitor the traffic in a highly dynamic network which consists of independent nodes that can join and leave the network dynamically. Detection of attacks becomes difficult in such infrastructures.
2. ***Resource availability***: This is a major issue in MANETs. Protecting a network against specific threats and attacks and providing secure communication in such a dynamic environment requires various security schemes. Collaborative ad hoc environments also allow implementation of self-organized security mechanisms.
3. ***Scalability***: Nodes in MANETs may join and leave the network at any time. Due to this mobility, the network topology and the number of nodes in an ad hoc network keeps changing. Thus, scalability is a major security issue and the security mechanisms should be able to handle both large and small networks.
4. ***Cooperativeness***: One of the requirement in routing algorithms for MANETs is that nodes should be cooperative and non-malicious. A malicious node may fail to corporate and disrupt the network operation by changing the routing information. This results in a malicious attack that will disrupt network operations by protocol specifications.
5. ***Dynamic topology***: Changeable node memberships may disturb the trust relationships among nodes. The trust may also be disturbed if some nodes are compromised. This dynamic behavior could be better protected with distributed and adaptive security mechanisms.
6. ***Limited power supply***: The nodes in mobile ad hoc networks are battery operated and their restricted power supplies must be considered. A node in MANET may start behaving in a selfish manner as soon as it discovers that has only a minimum battery power or limited electricity supply.
7. ***Bandwidth constraint***: In MANETs, variable low capacity links mean they are more susceptible to external noise, interference and signal attenuation effects than wired networks.
8. ***Adversary inside network***: MANETs allow mobile nodes to freely join and leave the network at any time. It is difficult to detect when a compromised node starts behaving maliciously. Such

compromised or malicious node attacks are considered more severe than external attacks.

9. ***No predefined boundary***: In MANETs, precise definition of a physical boundary of the network is not possible. Nodes work in a nomadic environment where they are allowed to join and leave a network at any time. As soon as a node enters the vicinity of a particular node in MANET, it starts communicating with that node and can perform attacks such as eavesdropping, impersonation, replay and denial of service (DoS).

5.3 Classification of Security Attacks

MANETs suffer from lack of security and are exposed to different kinds of attacks. The most challenging issue is securing the shared wireless medium. The foremost step in developing security schemes to counter such attacks is knowing that an attack can happen. Due to lack of central coordination and the sharing feature of a wireless medium MANETs are more susceptible to different forms of attacks than wired networks. The security attacks on MANETs can be broadly classified as shown in Figure. 5.1 and the descriptions of the attacks are explained below.

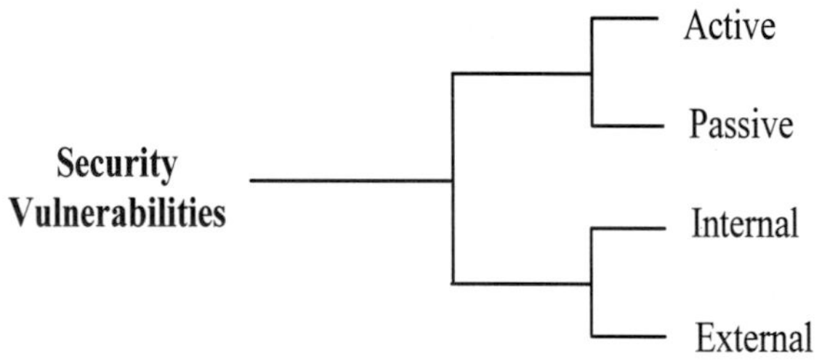

FIGURE 5.1
Different Types of Attacks on MANETs

1. **Active attacks** *alter*, *inject*, *delete* or *destroy* the data being exchanged in a network. The main intention of such type attacks is to damage the network or disrupt the operations. *Example: fabrication or masquerading attacks, message modifications, message replays, and DOS attacks.*

2. **Passive attacks** are intended to *learn* or *make use of information.* They have no plan to damage the network operations as the

contents of the packets are not modified in such attacks. Also, passive attacks do not affect the network resources. Detection of passive attacks is very difficult because no variation in the operation of the network can be observed. *Example: eavesdropping, Release of message contents and traffic analysis.*

A detailed classification of different active and passive attacks on MANETs is given in Figure 5.2.

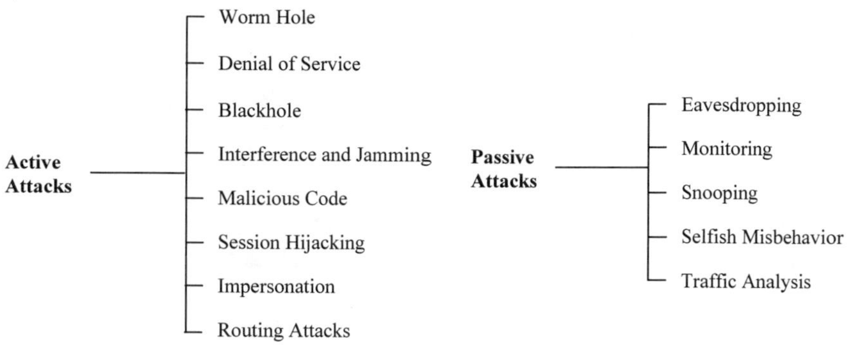

FIGURE 5.2
Classification of Active and Passive Attacks

Based on the locations of the nodes in relation to the network, two more types of attacks are *internal* and *external* and they are defined as follows:

1. **External Attacks** are carried out by nodes or group of nodes that do not belong to the network. Such attacks send fake packets to interrupt the ongoing operation and performance of the network. External attacks try to cause *congestion* in the network by advertising wrong routing information and denial of services (DoS).
2. **Internal Attacks** are carried out by nodes or groups of nodes that are part of the network. Since the attackers are within the network, the wrong routing information generated by these compromised or malicious nodes is difficult to detect. Internal attacks are hard to detect and more severe than external attacks.

5.4 Security Threats at Different Layers

Attacks can come from all directions and target any node in a network. In this section, categorization of security attacks in different layers of the protocol

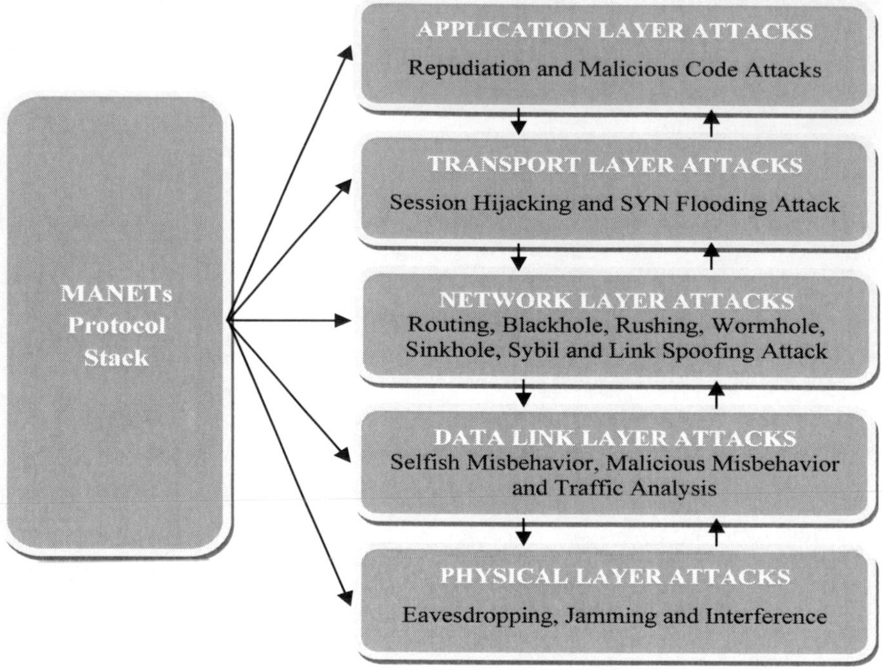

FIGURE 5.3
Security Threats at Different Layers in MANETs

stack is discussed and depicted in Figure 5.3. The figure lists the different attacks at each layer of MANETs. The following sections explain the security attacks on different layers of MANETs.

5.4.1 Physical Layer

The attacks on the physical layer cannot be performed independently. They require help from the hardware resources. Security vulnerabilities on the physical layer are explained below.

1. **Eavesdropping**
   Eavesdropping can be defined as secretly listening to a conversation. It is a process of interception and reading of information from a network by snooping on the transmitted data. The interception of the wireless communication media can easily by unintended receivers be done easily by the receiver who tunes to the proper ongoing communication frequency. The purpose of the attack is to gain access to confidential information such as public key, private key, location and passwords of the nodes, that should be kept secret during communication. Figure 5.4 shows the evesdropping scenario

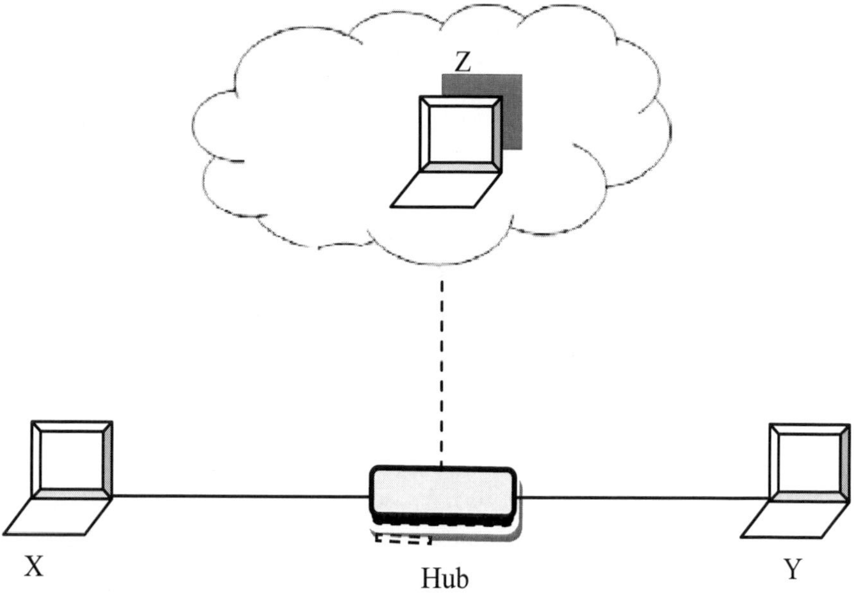

FIGURE 5.4
Eavesdropping

in a MANET. Nodes X and Y communicate with each other by exchanging data through the hub. Unknown to X and Y, node Z is listening to all the transmitted traffic in the network. Z intercepts the communication to hear or gather the data exchanged between node X and node Y. Z may also be a malicious hacker who can use snooping techniques to capture *passwords, monitor key, strokes* and even *gain access to login information.*

2. **Jamming and Active Interference**
Jamming is a special category of DoS attack in which a radio signal can be jammed or interrupted, causing the message to be corrupted or lost. The attacker initially monitors the wireless medium to determine the frequency at which the destination node is receiving signals from the sender and than transmits signals of the same frequency to send data to the receiver, thereby disrupting communications. A jammer can use many strategies to interfere with other wireless communications and few are listed below.

Constant Jammer: A constant jammer continuously emits a radio signal that represents random bits.

Deceptive Jammer: These jammers transmit semi-valid packets. The packet header is valid in a semi-valid packet but the payload is useless.

A random Jammer: Random Jammer alternates between *sleeping* and *jamming* the channel. In sleeping mode, the jammer jams for a random period, behaving like a constant jammer or deceptive jammer and then turns its transmitters off for another random period.

Reactive Jammer: A reactive jammer will become active whenever it senses a transmission and it targets the receiver. The main aim of the reactive jammer is to introduce noise in the packet by modifying the bits in the packet so that when a checksum is calculated at the receiver end, the packet is discarded.

Interference: The main motive in interference is to decrease the signal-to-noise ratio (SNR) of a received signal. Such an attack is accomplished by interfering with the original signal. This attack is performed by introducing noise signals of the same frequency range as used in the communication.

Stolen or compromised attack: These kinds of attacks occur from a compromised entity or stolen device. An attacker may gain access to a stolen device that equates to physical capture of a node which may occur due to device tampering.

5.4.2 Data Link layer

Several algorithms used in the data link layers of MANETs are vulnerable to DoS attack. Security attacks performed on the link layer of mainly affect the cooperation of the link layer protocols among the immediate upper (network) and lower (physical) layers of the protocol stack. The attacks cause link breakage which in turn initiates a route rediscovery phase, route discovery failure, energy consumption and so on. The attacks performed on the data link layer are as follows:

1. **Selfish misbehavior of nodes**
   Attacks by misbehaving (selfish) nodes degrade the performance of other nodes by their failure to cooperate with and disruption of the network. Selfish nodes may fail to forward packets to other nodes or may drop packets intentionally to conserve their resources such as battery power. Such attacks in MANETs produce congestion and may also prevent end-to-end communications among nodes, especially if the selfish node occupies a critical point in the path. Retransmissions of dropped data packets will reduce network performance.

2. **Malicious behavior of nodes**
   The main objective of malicious nodes is to disrupt the operations of routing protocols. By advertising wrong routing information, these malicious nodes receive information intended for other nodes in the

network. The impact is high when communication occurs between neighboring nodes.

3. **Denial of service (DoS)**
DoS attacks generate malicious actions with the help of compromised nodes and cause to security risks. Moreover, it is difficult to detect the wrong routing caused by compromised nodes. Such a compromised route resembles a normal route and leads to severe problems. For example, a compromised node involved in communication may drop packets in turn leading to degradation in the QoS offered by network.

4. **Attacks on network integrity** To provide secure communication and QoS, network integrity is an important requirement. There are various threats which exploit the vulnerabilities in routing protocols so as to introduce the wrong routing information in a network.

5. **Misdirecting traffic**
A malicious node advertises a fake routing request so that other nodes direct route replies to the malicious node that receives information intended for the owner of the address.

6. **Attacking neighbor sensing protocols**
Malicious nodes advertise fake error messages that mark important link interfaces as broken, resulting in decreased throughput and QoS of the network.

7. **Traffic analysis**
Traffic analysis intends to analyze the traffic flow between the nodes in a network. First the attacker reveals information regarding the sources and destinations of communicating entities, network topology, location of nodes, communication and responsibilities of nodes, and confidential information about network topology. The revealed information is then used to carry out further attacks.

5.4.3 Network Layer

The nodes in MANETS act as both hosts and routers. They perform router functions by finding and maintaining routes to other nodes in the network. Network layer protocols enable hop-by-hop connections of nodes. The purpose of a routing protocol is establishing an optimal and efficient route between the source and destination nodes. All individual nodes in MANETs can decide which neighbor nodes to choose to forward data packets. Network layer nodes attack by placing themselves in the active path between the source and destination nodes during the routing phase; thus, it is easy for malicious nodes to attacks in such an environment. These attacks disrupt communications throughout an entire network. Some of the most common network layer attacks are described below.

Routing Attack

An attacker places itself in the path between the source and destination nodes and is able to absorb and control the network traffic flow between the communication entities. Different routing attacks in MANETs are shown in Figure 5.5. In Figure 5.5 Nodes A,B, C, and D are involved in a communication where source node A is sending packets to node D. A malicious node Z could place itself into the routing path between sender A and node B (Figure 5.5b). By placing itself in the communication path, node Z diverts the data packets exchanged between A and D, which results in significant end-to-end delay between source and the destination. Further, the malicious node Z can also create routing loops, thereby disrupting route discovery as shown in Figure 5.5c. Figure 5.5d is a special case of routing attack in which the malicious node Z overflows the routing table by creating enough routes to prevent new routes from being created.

Blackhole Attack

A malicious node claims that it has an optimum route to the destination node. Once the malicious node introduces itself between the communicating nodes, it intends to absorb all the data by false replies to the route requests without having an active route to the destination. If the attacker is closer to the destination, blackhole attack causes heavy damage to the network. Since this malicious node immediately replies to the RREQ message from the sender with RREP messages with highest sequence numbers to settle in the routing table of the victim, the requesting node assumes that route discovery complete and will ignore rest of the RREP messages; the sender node keeps sending data packets to malicious node which will drop the data packet. In this way the malicious node attacks all RREQ messages and takes over all routes. A black hole attack scenario is shown in Figure 5.6. Node A initiates route discovery to find a fresh path to the destination node D. A malicious node Z announces that it has an active route to the destination node D by sending a RREP packet in response to the RREQ packets from A. When node A receives the RREP packet from malicious node Z, it assumes that there is an active route to the destination node D through node Z. It ignores all other $RREP$packets and begins its data communication which results in all the data packet being consumed or lost at malicious node Z.

Rushing Attack

A malicious node claims that it has an optimum route to the node whose packets it wants to intercept. An attacker, on receiving the $RREQ$ packet, quickly floods the $RREQ$ packet throughout the network before other nodes that receive the same RREQ can react. Rushing attacks are mainly against the on-demand routing protocols and are extremely difficult to detect.

In Figure 5.7, source node A and destination node H are connected wirelessly and node Z represents the rushing attack node. Node Z performs the attack by quickly broadcasting the RREQ messages to ensure that the RREQ message from itself reach earlier than the requests from other nodes. This

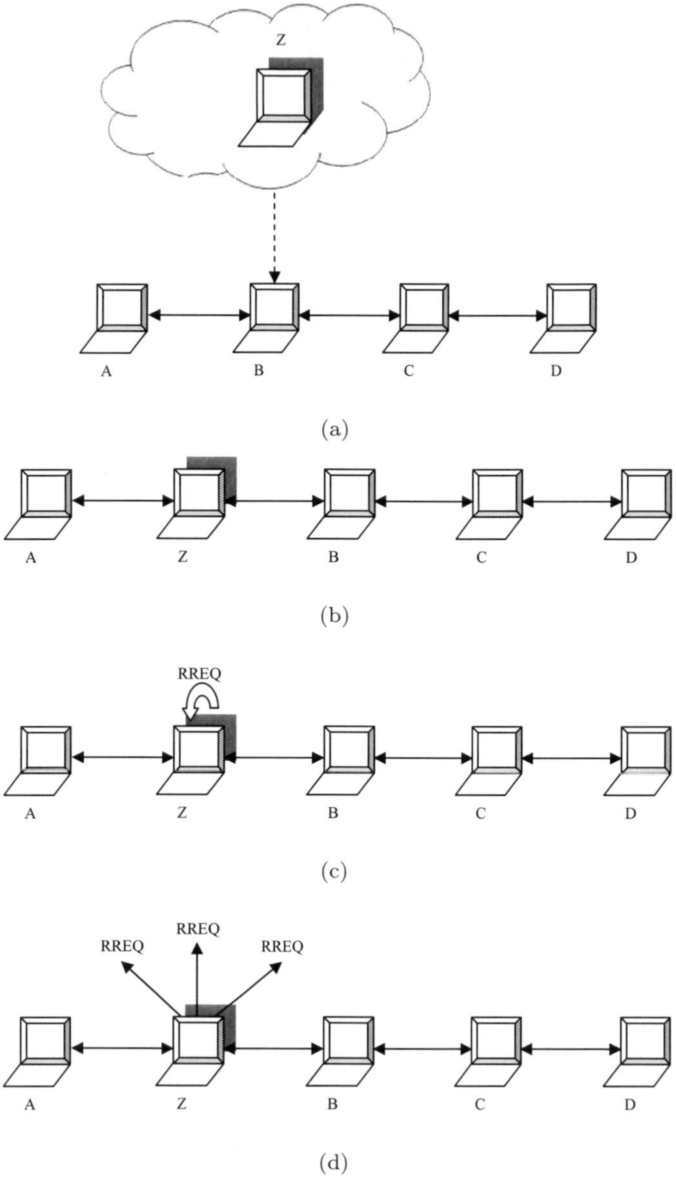

(a) Malicious node trying to introduce itself In the path.
(b) Malicious node participating in the path.
(c) Malicious node creating routing loops.
(d) Malicious node overflowing routing tables.

FIGURE 5.5
Routing Attacks

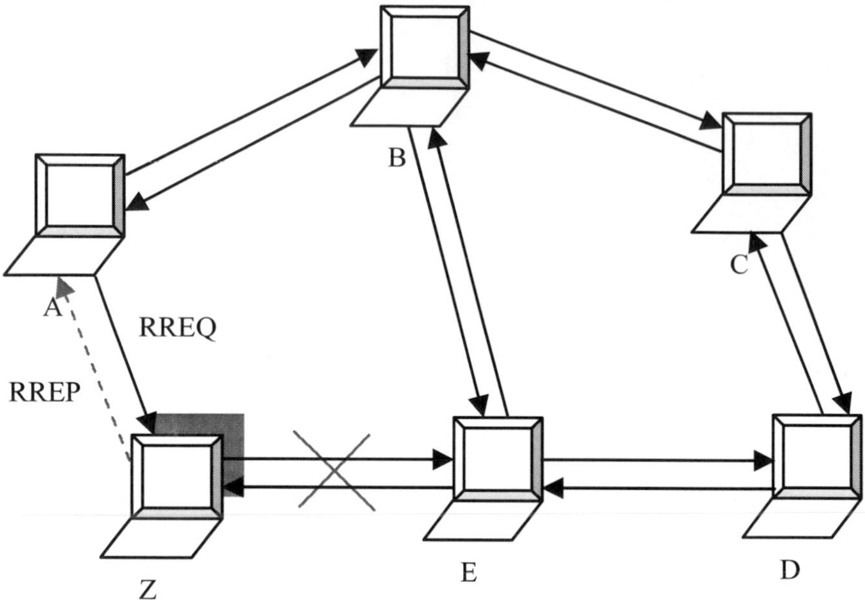

FIGURE 5.6
Blackhole Attack

results in discarding the RREQ from source *A* by the nodes *F* and *G* as duplicate RREQ packets. Thus, in the presence of rushing attack, source node *A* is unable to find safe route to its destination.

Wormhole Attack
Wormhole is performed by two attackers positioned in the network. A tunnel between two malicious nodes is referred to as a wormhole. One malicious node receives data packets in the network and tunnels them to the other malicious node. Wormhole attacks pose severe threats to MANET routing protocols. Attackers use wormholes to make their nodes appear eye-catching so that more data is routed through their nodes. The gravity of this attack lies in its ability to be launched against all communications that provide authenticity and confidentiality. For example as shown in Figure 5.8, source node *A* broadcasts an RREQ to find a route to a destination node *J*. The RREQ is forwarded to *A* neighbors *B* and *D* in a regular fashion. Nodes *Z1* and *Z2* are two colluding attackers that target node *A* to be attacked. During the attack, however, node *Z1*, which received the RREQ forwarded by node *B* records and tunnels the RREQ to its partner node *Z2*. Upon receiving this RREQ, node *Z2* rebroadcasts it to its neighbor *C*. This RREQ is able to reach the destination node *J* before the rest of the RREQ packets since this RREQ packet is passed through a high speed channel. Therefore, node *J* selects route *J-C-B-A* to unicast an

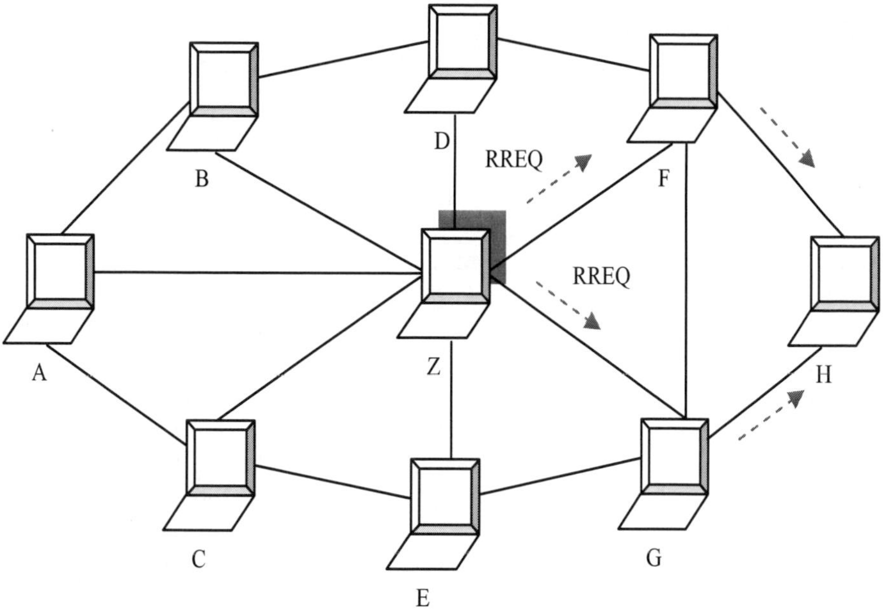

FIGURE 5.7
Rushing Attack

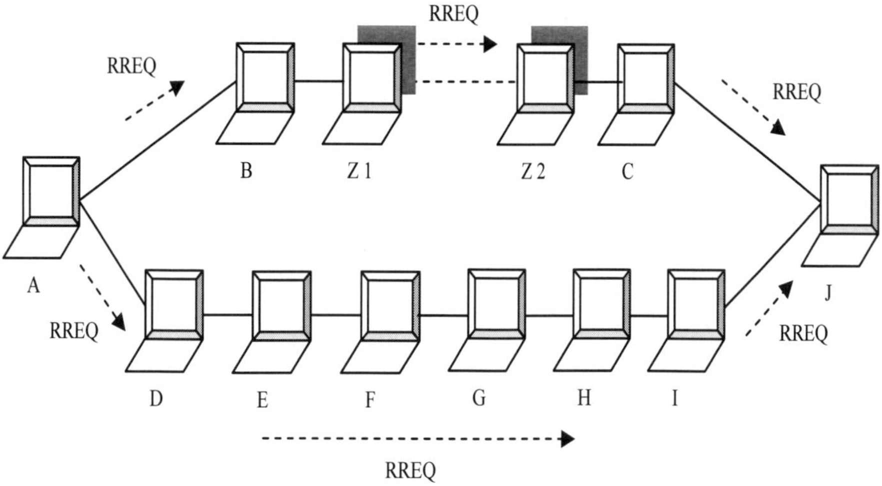

FIGURE 5.8
Wormhole Attack

RREP to the source node *A* and ignores any other RREQ that arrived later. As a result, node *A* starts forwarding its data on route *A-B-C-J* that passes through *Z1* and *Z2*.

Sinkhole Attack

A compromised node or malicious node poses itself as a specific node by broadcasting wrong routing information and receiving all network traffic. In other words, the attacking node tries to offer a very attractive link that attracts the data from its neighboring nodes which cause traffic to bypass this node as shown in Figure 5.9. A simple traffic analysis as discussed earlier could be carried out. Other forms of attacks could be combined such as selective forwarding or DoS. Sinkhole attacks performed in AODV protocol tend to maximize the sequence number or minimize the hop count. The path presented through the malicious node appears to be the optimal route for the nodes to communicate. Sinkhole attack is considered one of the most severe forms of attack in wireless ad hoc networks. In DSR protocol, a sinkhole attack modifies sequence numbera in RREQs.

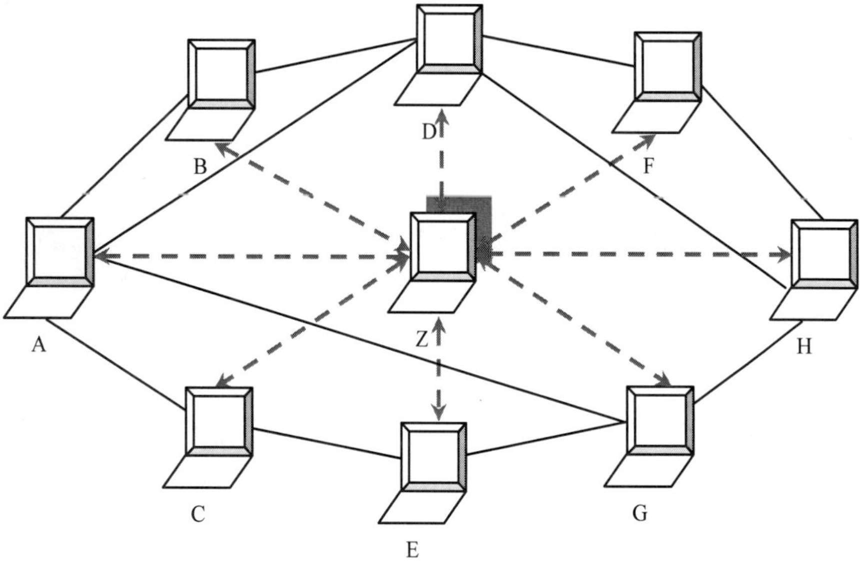

FIGURE 5.9
Sinkhole Attack

Link Withholding Attack

A malicious node does not advertise the links of specific nodes or a group of nodes. This deliberate action results in link loss to these nodes. This type of attack is severe in the OLSR protocol.

Link Spoofing Attack
A malicious node broadcasts fake routing information to its non-neighbors to disrupt routing. Malicious node Z in Figure 5.10 specifically manipulates the data or routing traffic which causes source node A to make a wrong decision about the fake link provided by node Z.

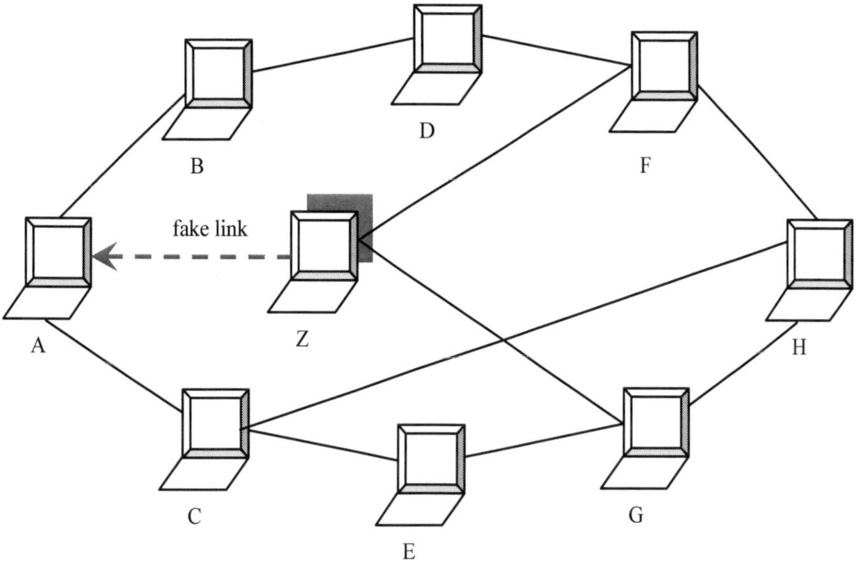

FIGURE 5.10
Link Spoofing Attack

Replay Attack
In MANETs, the network topology keeps changing frequently due to the mobility of nodes. Thus, a malicious node which is able to record control messages of other nodes performs a replay attack by resending the captured sequence to the destination later. This results in other nodes loading their routing tables with stale routes. These replay attacks are later misused to impersonate a specific node or disturb the routing operation in MANETs.

Resource Consumption Attack
A compromised node tries to consume battery life by requesting repeated route discovery or forwarding unnecessary packets to the victim. Such attacks are also known as sleep deprivation attacks performed against devices that have no services to offer within the network.

Sybil Attack
A single malicious node poses itself as a large number of nodes. Sybil generates fake identities to represent multiple identities for a malicious node. The additional identities are called Sybil nodes. A sybil node may formulate a new

identity for itself or may steal an identity of a legitimate node. The presence of sybil nodes in a network makes it difficult to recognize a misbehaving node and these forms of attacks prevent fair resource allocation in the network. Moreover, in certain applications, sensors performing voting for decision making find disparity in the voting outcome due to the presence of these duplicate identities. Thus, the sybil nodes are able to disrupt the normal operation of routing protocols in a network by appearing at various locations. As shown in Figure 5.11, node *S* is connected with nodes *A, B, D* , and *F* and the malicious node *Z1*. If *Z1* represents other sybil nodes as *Z2, Z3*, and *Z4* (using corresponding secret keys), *S* will believe it has eight neighbors instead of five.

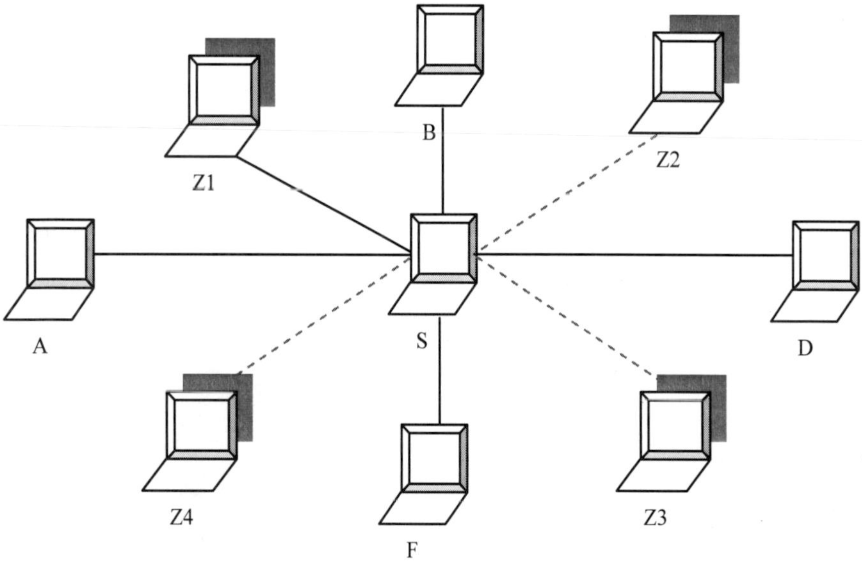

FIGURE 5.11
Sybil Attack

5.4.4 Transport Layer

In MANETs, specific functionalities at the transport layer include establishing end-to-end connection, reliable packet delivery, flow control and congestion control. Similar to TCP protocols in the Internet, the transport layer in MANETs is also vulnerable to the session hijacking attack and SYN flooding attack which are discussed below.

Session Hijacking
In a session, all the authentication will be carried out before the commencement of the session. Session hijacking takes advantage of the fact that most

communications are protected only at session setup and not subsequently. The adversary spoofs the IP address of the destination node and masquerades as one of the end nodes of the session, thus hijacking the session as a legitimate system.

SYN Flooding Attack
SYN flooding is a DoS attack in which an attacker establishes a large number of half-opened TCP connections with a victim node. The attacker never completes the handshake to fully open the TCP connection between two communicating parties. The attacker begins to send a large number of SYN packets to a victim node, spoofing the return address of the SYN packets. The victim node on receiving the SYN packets, stores them in a fixed-size table and sends SYN-ACK packets to the sender and keeps waiting for ACK packets. These pending connection requests run over the buffer and make the system unavailable for long time. The establishment of a TCP connection between two communicating parties is done by completing a three-way handshake as described in Figure 5.12.

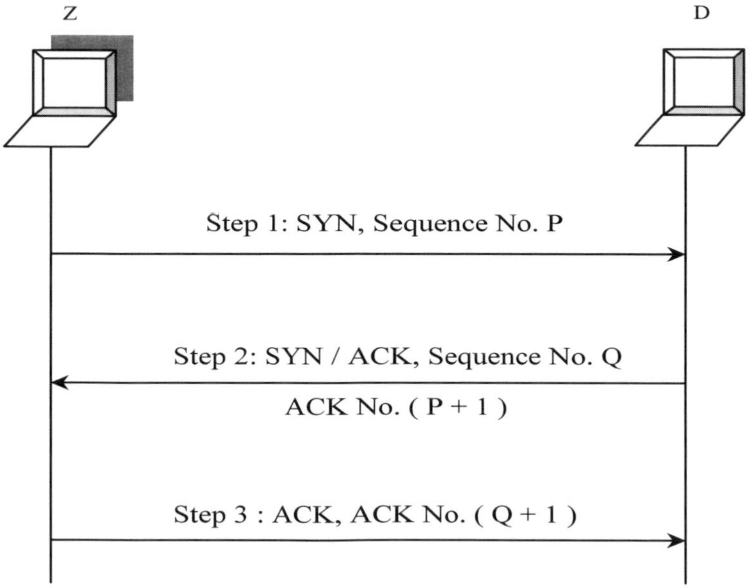

FIGURE 5.12
SYN Flooding Attack

In SYN flooding attack, node *Z* initiates a large number of TCP connections with the victim node *D*. Node *Z* spoofs the return address of the SYN packets and does not complete step 3 of these TCP connections. Node *D*, upon receiving the SYN packet from the attacker, immediately issues the SYN-ACK packets to the spoofed address, which often does not exist. *D* awaits reception of ACK packets (in step 3). A large number of these half-opened connections may overflow the buffer maintained by D and will not allow node D to accept any other connection request.

5.4.5 Application Layer

End user applications are accessed by the users with the help of the application layer. It supports many protocols such as HTTP, SMTP, TELNET, and FTP and also contains user data that are prone to attacks and access by attackers. Various end user applications require connection with storage devices that are prone to many viruses. Possible security attacks at this layer are mobile viruses, worm attacks and repudiation attacks. Application layer protocols are also vulnerable to many DoS attacks.

Malicious Code Attacks
Malicious code attacks include viruses, worms, spyware and trojan horses. Both operating system and end user application can be infected by such attacks. These malicious programs can spread themselves through the network and cause the entire system and network to slow down. One effective way to prevent various attacks is to implement a firewall as well as use an intrusion detection system (IDS) to prevent gaining unauthorized access to a service.

Repudiation Attack
Repudiation refers to a denial or attempted denial of participation in all or part of a communication by node.

5.5 Security Schemes

This section covers security schemes such as active and passive attack prevention, intrusion detection, and response approaches in depth. The states of the art of well known security algorithms will be discussed in detail.

5.5.1 Active Attack Prevention Approaches

The approaches to MANETs security can be classified as active or passive and they are described below.

In ad hoc networks, as data are forwarded in multi-hop fashion to a destination, node security is vital because the selfish nature of nodes may make them vulnerable to external attacks. Malicious nodes may deny services to legitimate nodes, cause DoS, and suffer routing attacks. Nodes in MANETs have limited processing power and short battery lives and thus must conserve resources. For these reasons, the authentication and encryption methods utilized in wired networks cannot be applied to MANETs even though MANETs require these two functions to provide security to end users and protect protocols in the stack [11] [16].

Steiner et al. [77] proposed a solution known as the group key Diffie-Hellman (GDH) model. The mobile certification authority (MOCA) protocol based on the public key infrastructure (PKI) was developed by Yiet al. [72]. The protocol manages heterogeneous mobile nodes in MANETs. Poor key management leads to DoS attacks and Capkun et al. [55] developed the maximum degree algorithm (MDA) as a solution.

Routing in MANETs is a challenge because the nodes in ad hoc networks because the nodes perform routing functions and security protocols are not included in routing protocol designs. Routing tables serve as the bases of all network operations. Corruption of a routing table produces severe consequences.

MANETs utilize two well known protocols for routing: the ad hoc on-demand distance vector (AODY) and the destination sequenced distance vector (DSDV). The AODV vector is reactive in the sense that it finds a route from source to destination only when required. DSDV is proactive; it maintains a dynamic routing table at all times.

Self-Organized Security (SOS)

Yang et al. [33] proposed a reactive routing based on AODV that utilizes a unified network layer prevention scheme known SOS. The self-organized approach exploits fully localized design and does not any a priori, trust, or secret associations between nodes.

A node must possess a token to participate in network operations. Its neighbor nodes continuously monitor it to detect any misbehaviors in its packet forwarding functions. A node must renew its token on expiration through its neighbors. The validity period of a token is defined as the time it has remained in the network and performed well. Over time, a node accumulates performance credit and must renew its token less frequently.

Techniques for Intrusion-Resistant Ad Hoc Routing Algorithms (TIARA)

Technique for Intrusion-Resistant Ad Hoc Routing Algorithm (TIARA) Ramanujam et al. [1] proposed a new approach called TIARA for building intrusion resistant ad hoc networks using wireless router extensions. TIARA is reactive by extending the capabilities of ad hoc routing algorithms. It is able to detect and eliminate DoS attacks and efficiently handles intruders without modifying existing routing algorithms. The authors also describe a new network mechanism for detection of and recovery from intruder-induced malicious faults.

Secure Efficient Ad hoc Distance vector (SEAD)

SEAD is a DSDV-based secure routing scheme proposed by Hu et. al. [70] which uses an efficient one-way hash function. SEAD is designed to guard against DoS attacks that cause nodes to consume excess bandwidth or processing time. Since nodes have limited CPU processing ability, SEAD does not use symmetric cryptographic operations.

Awerbuch Technique

Awerbuch et al. [10] carried out a study on the behaviors of routers in the presence of Byzantine faults. They used an on-demand secure routing protocol (OSRP) in which they defined the reliability metric based on past records, then use the metric to choose a secure path. The reliability metric is interpreted through a list of link weights. Any link carrying a high weight has low reliability. Each node in a network maintains a weight list that is updated dynamically whenever a fault is detected. Asecure adaptive probing feature embedded in the normal packet stream identifies faulty links.

Key Management Service (KMS)

Zhou and Hass [75] devised KMS as a solution to the problems of AODV and DSDV routing. KMS is a hybrid approach combining AODV and DSDV to defend against DoS attacks on routing in ad hoc networks by utilizing the benefits of multiple routes between nodes. The authors stress the need to handle outdated routing information due to the dynamic changes in topology that are analogous the compromised routing attacks. Even with the existence of compromised nodes within the network, the routing protocol finds the routes with adequate numbers of correctly working nodes. If a fault occurs in the primary route, the protocol is able to switch to an alternate route and continue the routing process. This protocol makes use of replication and new cryptographic methods such as threshold cryptography to deliver a secure key management service which is considered the core of a security framework.

Secure Routing Protocol (SRP)

Papadimitratos et. al. [48] proposed a secure routing protocol (SRP) to guarantee correct route discovery. Fabricated, compromised and replayed route replies are discarded. SRP never allows fabricated and other forms of route replies to reach the route requester. The strength of SRP in route discovery is that no intermediate nodes need to be trusted, assuming that a security association exists between the end points of a path. This is done by allowing a unique random query identifier to reach the destination. At the destination, a message authentication code is computed over the path and a route reply is constructed and returned to the source to grant the request.

Ubiquitous and Robust Authentication Services

Luo et. al. [42] propose a different preventive solution for DoS attack called ubiquitous and robust authentication services for ad hoc wireless networks. The principle is to dispense the functionality of authentication servers in each node. The result is that each node in the network is able to collaboratively self-secure itself by using the certificate-based approach. Centralized management is minimized. Such an approach supports ubiquitous security for mobile nodes and scales the network to varying size and robustness against adversary link breaks. A complete group of fully distributed and localized protocols is provided to make practical deployment feasible. Communication efficiency is key to conserving the wireless channel bandwidth and the protocol features independently from both the underlying transport layer protocols and the network layer routing protocols.

Alliance of Remote Instructional Authoring and Distributed Networks for Europe (ARIADNE)

The alliance of remote instructional authoring and distributed networks for Europe (ARIADNE) is another important secure on-demand routing protocol developed by Hu et. al. [69]. The design of ARIADNE is based on the DSR algorithm and follows symmetric cryptography which prevents attackers from tampering with routes and nodes. ARIADNE performs like DSR without optimization. The protocol consists of three stages. In the first stage, the authors present a method that enables the target to verify the authenticity of the route request. The second stage is concerned with authenticating data in route requests and route replies. The mechanism includes a key management protocol (with synchronized clocks), digital signatures, and standard MAC (message authentication code). In the third stage, the algorithm runs to ensure that no node is omitted from the node list in the request by running an efficient per-hop hashing technique.

Marti Technique

A variation of DSR was proposed by Marti et. al. [56] to mitigate routing misbehavior in MANETs. This method outperforms DSR in increased throughput. The authors complement the DSR protocol by close detection of denied packet forwarding and setting up a path rater for trust management and rating routing policy that every path uses. Such mechanisms enable nodes to avoid malicious nodes in their routes working thereby, providing a detective and reactive protection measure. The protocol includes the nodes with better paths to enhance throughput without including non-cooperative malicious nodes in routing process.

Security-Aware Ad hoc Routing (SAR)

Yi et. al. [72] integrate security attributes in ad hoc route discovery and recommend a new routing method for MANETs called security-aware ad hoc routing (SAR). SAR implements security to improve the relevance of the routes discovered by ad hoc routing protocols. Nodes in ad hoc routing protocols communicate with their neighbors through route request (RREQ) packets and route reply (RREP) packets. The security metrics are embedded into RREQ packets. When these intermediate nodes receive the packets, depending on the security level of the intermediate nodes, they nodes are able to process or forward the packets. RREQ packets that have not attained a necessary security level are dropped or the source may receive the sent RREP packets from the destination or the intermediate nodes. Such a scheme is helpful for preventing attacks.

The techniques surveyed address issues such as DoS, authentication, selfish nodes and routing protocol attacks. MANETs assume that each node allows other nodes to be acquainted with their presence and willingness to participate. A mobile node broadcasts a beacon signal to let other nodes know of its presence.

Binkley Technique

Brinkley et al. [15] proposed a method to reduce DoS threats such as replay attacks in an address resolution protocol (ARP) or an ad hoc routing spoof that destroys a link layer route to a host. Their protocol addresses link security. On receiving transmitted beacons, a host first authenticates them. If the beacons are authentic, the nodes add the MAD-to-IP address binding to the beacon; the information is added to the table of authentic bindings.

Kong and Luo Technique

Kong et. al. [35] and Luo et. al. [42] propose other security schemes by implementing a threshold secret sharing mechanism which supports ubiquitous security services for mobile hosts. This system employs localized certification schemes to enable ubiquitous services. The implementation is based on RSA cryptography that offers distributed localized certificate services such as issuing, renewal, and revocation. The concept of threshold secret sharing and updating each entity's secret share periodically is done to enhance robustness against break-ins. The system is robust against break-ins and scales to network size. The threshold secret sharing mechanism enables each entity to hold a secret share and multiple entities in a local neighborhood in cooperation provide a complete set of services.

This concludes the brief description of the active attack prevention approaches for MANETs. The taxonomy of the active prevention approaches is given in Figure. 5.13.

5.5.2 Passive Attack Prevention Approaches

One of the significant challenges in MANETs is the identification of selfish nodes. These nodes do not perform their assigned roles and lead to denial of service, congestion, and lower throughput. In a MANET, many nodes try to conserve battery life and consequently exhibit selfish behavior by dropping packets rather than forwarding them. A selfish node will not participate in packet forwarding and will always try to conserve energy.

Buttyan Technique

Buttyan et. al. [18] showed through simulation that 80% of the energy in nodes is spent in forwarding packets. The authors also introduce a special counter named nuglet that tracks the selfish behaviors of nodes.

CORE

In an effort to solve the problem of selfish nodes, Michiardi et. al. [46] proposed a collaborative reputation model (CORE) in which every node monitors

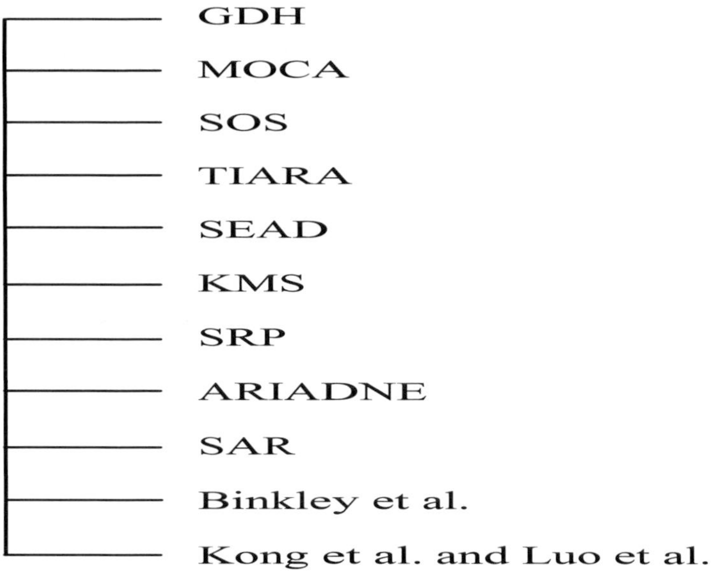

FIGURE 5.13
Active Attack Prevention Approaches

the behavior of the neighboring nodes for a requested function. CORE also aggregates data regarding the execution of the intended function. If there is a match in the observed result and the expected result of the function, a positive value is assigned to the observation. The advantage of this scheme is to enable a node to identify whether any of its neighbors are selfish and gradually isolate such nodes.

CONFIDANT

Cooperation of nodes: fairness in distributed ad hoc networks (CONFIDANT) is a network layer protocol proposed by Buchegger et al. [17] to promote fairness in wireless ad hoc networks. CONFIDANT is a reputation-based trust protocol for dealing with vulnerabilities posed by selfish nodes in MANETs. These misbehaving nodes experience Byzantine failures caused by malicious activities or component failures. In CONFIDANT, each node maintains reputation indices for its neighbors based on their behaviors. The indices are used to isolate misbehaving nodes.

Guardian Angel

Avoine et. al. [8] propose a model called guardian angel which is appropriate for networks of communicating devices with security modules. Guardian Aangel is a cryptography-based fair key exchange model that uses a probabilistic approach without a trusted third party in key exchange.

This concludes the brief description of the passive attack prevention approaches for MANETs. The taxonomy of the passive prevention approaches is given in Figure. 5.14.

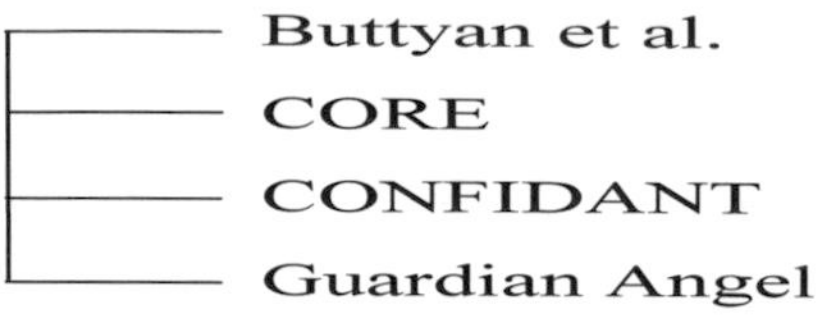

FIGURE 5.14
Passive Attack Prevention Approaches

5.5.3 Intrusion Detection and Response Security Approaches

Kachirski and Guha Technique

Kachirski and Guha in [37] propose an effective intrusion detection scheme using multiple sensors in wireless ad hoc networks. The implementation is an efficient and bandwidth-conscious intrusion detection system which targets intrusion at multiple levels. The sensors are organized in clusters which make the network more efficient and the intrusion detection agent (IDA) on a cluster head employs independent detection and decision making upon gathering information from other nodes. This scheme uses mobile agents for communication with multiple types of audit data. The system is flexible in supporting new types of audit data by incorporating extra agents which monitor the new audit data. framework is made more efficient by limiting resource consumption to a certain number of nodes. The scheme can be expanded easily for multi-layer mobile networks. It can also be applied in systems that require medium security and high efficiency.

Zhang and Lee Technique

Zhang and Lee [71] examined the vulnerabilities of wireless networks to demonstrate why an intrusion detection system (IDS) must be included in the security architecture of a mobile computing environment. They developed and evaluated a key method incorporating anomaly detection in collaborative decision making in mobile ad hoc networks. The authors indicate that individual intrusion detection (ID) agents can work independently and collaborate in decision making. Each ID agent running on a node monitors local activities. If a node decides that intrusion has occurred, it collaborates by initiating an alarm response. It then conducts an investigation via a distributed consensus algorithm to determine whether the evidence is it received is strong enough to detect an intrusion. This algorithm fits the distributed nature of mobile networks and also works with multiple types of audit data. The flexibility to support multiple types of audit data is provided by adding additional data collection module in the ID an agent. Data mining is incorporated as a local intrusion detection method which distinguishes the intrusion detection rate and the false alarm rate.

Cooperative Intrusion Detection System

A cooperative intrusion detection system for ad hoc networks [34] was proposed by Huang and Lee in which a cluster head is elected by a group of nodes in a neighborhood referred to as citizen nodes. The head node controls the operation and monitors the citizen nodes. Nodes that collect features transmit the features locally to the elected cluster head. The ID agent makes use of concepts such as anomaly detection and data mining as in [71]. This model limits resource usage for ID to a few nodes, thereby improving the efficiency and arrives at satisfactory detection rate in mobile networks. Such a framework is suitable in environments that demand medium security and high efficiency. Further, this scheme can be easily expanded for multi-layered mobile networks.

Intrusion Detection Architecture

Albers and Camp [6] devised a general IDA for security in ad hoc networks. Each node runs a local IDA to detect intrusions and uses a form of external data to confirm such detections. The nodes collaborate and communicate using mobile agents that provide a scalable architecture. Additional functionalities can be added by incorporating more mobile agents with new tasks. This architecture relies heavily on mobile agents so its main drawback is the computational complexity required to create and manage all the agents.

Sun Technique

Sun et al. [12] devised an IDA model that uses a collaboration mechanism with anomaly detection. Anetwork is organized into logical zones and each zone consists of a gateway node and individual nodes that possess ID agents to identify intrusion activities. The agent generates an alert message whenever it detects an intrusion. The functions of the gateway node are aggregation and correlation of the alerts generated by the relevant individual nodes. The system runs an algorithm that aggregates the alerts based on their similar attributes. Only gateway nodes have the ability to use alerts to initiate alarms. This architecture can be applied to systems that have high security requirements.

Huang Technique

Huang et al. [34] focus on techniques for automatically constructing anomaly detection models that are capable of detecting new or unknown attacks. The authors propose a new data mining method that makes use of cross-feature analysis to capture the inter-feature correlation patterns in normal traffic. These patterns will then be used as normal profiles to detect anomalies caused by attacks. The authors implemented their method with two well known ad hoc routing protocols, DSR and AODV, and evaluated the performance using the NS-2 simulator. The authors conclude that the anomaly detection models based on the proposed data mining method can effectively detect anomalies.

Tseng Technique

Tseng et al. [66] proposed an intrusion detection system that can detect attacks performed on the AODV routing protocol. This scheme is the first of its kind to apply specification-based detection techniques to detect attacks in ad hoc network routing. Further, the proposed IDS is based on a distributed network monitor architecture that traces AODV request and reply flows. The implementation results in effectively detecting most of the routing attacks on AODV with low overhead.

Neighborhood Watch

An IDS protocol named neighborhood watch proposed by Sowjanya and Shah [58] specifies the usage of two neighboring nodes. One node is specifically assigned the task of identifying packets that are not modified while traveling in the network by comparing the information in each packet at each hop. The authors propose two modes of operation in which the passive mode protects a single host and active mode collaboratively protects the nodes that belong to a cluster. The cluster head runs a voting algorithm to detect the possibility of intrusion in the system.

Puttini Technique

Puttini et al. proposed an IDS architecture [51] that considers information as input data from the management information base (MIB). This scheme uses mobile agents and incorporates collaborative decision making. The algorithm is able to detect attacks at multiple levels and is highly scalable. Issues that need to be considered are security, computational cost, and management problems related to mobile agents. The model is based on distributed features and is efficient in its use.

Brutch and Ko Technique

Brutch and Ko [16] proposed an IDS model which is a statistical anomaly detection algorithm. The working principle is that the scheme first assumes that the audit trail generated from a host must be transformed to a canonical audit trail (CAT) format. The CAT file is used to generate session vectors representing the activities of the user sessions. Anomaly scores are evaluated by analyzing the session vectors against specific types of intrusive activities. Finally, the system generates warning reports if the scores cross some threshold value.

Indra Protecting

A distributed intrusion detection scheme called indra was proposed by Janakiraman et al. [50]. This system is based on sharing information among trusted peers in a network and protecting the network from intrusion attempts. This scheme is proactive and works on peer-to-peer (P2P) methodology to provide security. A authors propose cross monitoring technique called neighborhood watch which is a very simple technique the hosts on the P2P network appear as an immune system. Each host distributes information to all the interested peers in the network regarding attempted attacks. The victimized node is able to collect such information and sound an alarm by notifying all its adjacent hosts. Thus the system is able to react proactively on reception of such information. If the alarm is sounded, all further attacks to the hosts in the network are taken care of since all the adjacent nodes have already been informed of such attacks from other hosts.

The taxonomy of intrusion detection and response security approaches is given in Figure. 5.15.

FIGURE 5.15
Intrusion Detection and Response Security Approaches

5.6 SI Solutions to Secure Routing

SI solutions for secure routing in MANETs have been broadly classified into two types namely, (1) nonhybrid approaches and (2) hybrid approaches and both discussed in detail in the following sections.

5.6.1 SI-based Non-Hybrid Approaches

ACO-Oriented IDS Approaches

Abadi and Jalali [4] developed an ACO-basedalgorithm called antNAG for detecting intrusions. They assumed that intruders usually attack by considering the multiple vulnerabilities of systems. AntNAG organizes a set of all possible attack scenarios into a network attack graph (NAG). Scanning tools find the possible vulnerabilities of a system.

Additional information such as intruders' goals, template exploitation, and connectivity of network hosts are used to generate a NAG. Ants construct a set of critical exploits iteratively by incrementally adding exploits until all attack scenarios are exposed. At each graph construction step, an ant chooses an arc within which to move based on the amount of pheromone associated with the exploit. AntNAG utilizes an iteration-based solution that is improved by local search. Global updating rules initially modify the pheromone concentration on each trail. The NAGs generated by this approach are very large and complex.

Lianying and Fengyu [40] proposed an Sl-based intrusion detection system that separates the IDS into independent detection units. The separation enhances system performance and reduces the numbers of wrong decisions and improper detection rates. The presence of pheromone minimizes security vulnerabilities and eliminates some needs for exchange of information that indirectly increases network load.

ACO for Detecting Origin of Attack

Fenet and Hassas [25] claim to have developed the first IDS architecture using the ant colony metaphor to find the source of an attack. Their distributed intrusion detection and response system makes use of mobile and static agents. Astatic agent (pheromone server) is installed on each host that needs to be protected. The server is intended to send alert-like messages throughout the network in the event of an intrusion. The system includes a watcher (core component) and a static agent component ofthe detection system installed on each host that monitors its processes and network connections.

The mobile agents are known as lymphocytes. They are the core components of the response segment of the architecture. Their objective is to search for pheromone traces in the network. On detection of pheromone trails, they converge to the threatened component and initiate appropriate defensive actions based on the type of attack. The ant colony strategy allows the system to detect intrusions quickly and efficiently.

IDReAM [47] utilizes agent mobility to detect intrusions and respond to network attacks. The intrusion detection mechanisms are based on the human immune system and the intrusion response module is based on the ACO paradigm.

Each node runs a mobile agent (MA) platform that hosts intrusion detection agents (IDAs) and/or intrusion response agents (IRAs). IDA sevaluate suspicion indices (SIns) and move randomly in the network. When a SIn crosses a specific threshold value, an appropriate amount of pheromone is diffused. When IRA sencounter the pheromone traces during their random walks, they follow the pheromones back to their source and initiate a response to the attack.

A similar approach to intrusion detection in wireless sensor networks (WSNs)was proposed by Banerjee et al. [13] [14] and named IDEAS. To locate the sources of intrusions, this system relies on agents embedded on each sensor. The sensors monitor their hosts, peers, and network traffic to detect possible attacks. The network of sensors form a graph on which ant-like agents are randomly placed on nodes and traverse from their current nodes to adjacent nodes when they encounter a maximum number of violations expressed as pheromone.

Chen et al. [21] targeted DoS attacks. Their IP trace-back method can track the sources of DoS attacks without relying on network routers to perform detection. The first step is placing uniform amounts of pheromone on each router and placing ants on the victim nodes. Byknowing the topology, the ants can discover routers in their neighborhoods. They can also evaluate the probability of router movement to the next node based on traffic flow. Pheromone evaluation is based on average network traffic. The procedure is repeated until the boundary routers of the network are reached.

An intrusive analysis model based on the designs of honeypots and ant colonies was proposed by Chang-Lung et al. [8] [19]. A honeypot is a trap equipped with many vulnerabilities to attract intruders. The model is designed to determine the features of attacks and behaviors of intruders before a system is damaged. It utilizes all the resources of a honeypot and a pheromone value that is proportional to the significance of each resource. The honeypot is configured so that pheromones for each resource are updated only when an intrusion or malicious behavior is detected. Agreater concentration of pheromone indicates repeated attempts to compromise a resource. The ACO is applied to analyze the attackers' habits and interests and trace the trails of attacks.

IDReAM by Foukia [47] is an intrusion detection and response system to network attacks with agent mobility. This scheme is said to have intrusion detection mechanisms adopted from the human immune system and the intrusion response module is based on the ACO paradigm. Each node in this scheme, runs a Mobile Agent (MA) platform that defines (hosts) Intrusion Detection Agents (IDA) and/or Intrusion Response Agents (IRA). IDAs evaluate Suspicion Index (SIn) based on the local status moving randomly in the network. When the SIn crosses a specific threshold value, an appropriate amount of pheromone is diffused. When the IRAs encounter the pheromone traces during random walk, IRAs on following them back to their source, initiate a response to the attack. A similar approach in intrusion detection as above was proposed by Banerjee et al.,[13] [14] called IDEAS for Wireless Sensor Networks (WSN) to locate the source of intrusions. This system relies on agents that are embedded on each sensor. The agents perform the task of monitoring their hosts, peers and network traffic to detect the possibility of attacks. The network of sensors forms a graph on which ant like agents are placed on nodes randomly and traverse it. Ants traverse from current node to the adjacent node when they encounter a maximum number of violations expressed as pheromone. Visualizing the agents as emotional ants enhances the searching to be more accurate and efficient.

Chen et.al., target mainly the Denial of Service (DoS) attacks in [21]. An IP trace-back method is proposed which is able to trace the source of DoS attacks without relying on network routers that perform the detection process. In this scheme, the first step is to place the same amount of pheromone on each router and also place the ants on the victim node(s). By knowing the topology information, the ants are able to discover routers in the same neighbourhood. Further, the ants are able to evaluate the probability to move to the next node based on the traffic flow. Pheromone evaluation is done based on an average amount of network traffic. Thus the procedure is repeated until the boundary routers belonging to the monitored network are reached.

An intrusive analysis model based on the design of honeypots and ant colony is proposed by Chang-Lung et al.,[19][8]. The honeypot is a basically a trap with many vulnerabilities that targets to attract the intruders. So, this scheme aims to extract conclusions about the features of attacks and also the behaviour of intruders before the damage happens to the actual system. This proposed scheme associates all the network resources of the honeypot with Soroush et al. [57] presented an innovative work applying ACO for intrusion detection, unlike other approaches that use intrusion response. Soroush's system, like existing ones, requires a pre-existing dataset for training. The basis of the concept is the ant-miner algorithm [49] inspired by the foraging behavior of ants by classifying numerical data into predefined classes. The ant-miner algorithm is intended to deal with large amounts of complex data that must be analyzed to detect intrusions.

ACO for Induction of Classification Rules

During pheromone updating, the pheromone is increased. This process is repeated iteratively until the system constructs a huge rule set that can be used in test sets to classify network connections as intrusive or normal.

Junbing et al. [36] devised an improved ant-miner-based classifier for intrusion detection. It utilizes multiple ant colonies in contrast with the ant-miner algorithm that uses only one. The ants deposit pheromone specific to the colony to which they belong and the deposit affects only the ants of that colony. Only the best rules are added to the rule set.

The FORK of Ramachadran et al. [52] is another IDSproposal based on a variation of the ant-miner algorithm. It works efficiently under the constraints of ad hoc networks. Resources of ad hoc networks are limited and it is possibility that some of their nodes may be unable to detect intrusions and thus intrusion detection tasks are delegated to other nodes. The winner nodes detect intrusive behaviors.

The literature covers several studies [2] [3] [7] of authors who attempted to combine the concepts of generic algorithms and ACOs for the induction of accurate fuzzy classification rules. Zadeh [38] applied fuzzy set theory effectively for intrusion detection and their scheme displayed strong detection (DR) and false alarm rate (FAR) percentages. Agravat et al. [5] propose a modified ant-miner for intrusion detection an explain the concept of fitness function. The fitness function generates many rules of the same pattern pertaining to similar sets of data. All the quality rules generated by the entire ant colony are stored unlike other algorithms that store only the best rule produced by each ant.

Particle Swarm Optimization(PSO)

Dozier et al. [23] [24] presented a system that may be associated with an IDS. The approach makes it possible to identify attacks that may be disregarded because they are perceived as normal traffic. The authors address the choice of manually looking for vulnerabilities in a security system or allowing intruders to do so. The system involves modules called red teams that emulate the behaviors of hackers. The red teams use PSG for intrusion detection, making the IDS more effective.

PSO for Induction of Classification Rules

Guolong et al. [32] proposed a PSO-based approach for rule learning in network intrusion detection. Each particle is treated as a network connection that represents a rule. The algorithm is recursive in nature, thereby creating a particle population from a training dataset. For each network connection (particle), the algorithm computes fitness and updates the Pbest and Gbest values that denote the velocity and the position values of that particle. PSO cannot be directly applied to network intrusion datasets as attributes acquire distinct values. The authors were able to eliminate this limitation by suggesting a new coding scheme which maps the distinct attribute values to non-negative integers. A higher detection rate of intrusion was achieved by Chang et al. [20] who incorporated an accurate fitness function to the system described above.

Ant Colony Clustring (ACC) IDS approaches

ANTIDS proposed by Ramos and Abraham [53] incorporates the LF algorithm with intrusion detection. ANTIDS utilizes a number of ant-like agents that help build clusters and pick up and drop objects as required. The algorithm executes efficiently because it allows individual ants to find clusters of objects adaptively. It eliminates the need for agents to have memories and thus requires minimal resources. The process of finding an optimal solution is improved because the ants tend to move to areas of greater interest.

Tang and Kwong [64] [65] observed that the amount of data used for intrusion detection is voluminous. They proposed a variation of the original LF algorithm to address its inherent inefficiencies. The original algorithm suffers from homogeneous cluster formation and intensively discriminates dissimilar objects. To improve its efficiency, two different types of pheromones are used to guide the ant-like agents toward clusters where they can deposit or pick up objects as required. The authors later proposed an integrated multi-agent IDSarchitecture for industrial control systems [63].

The taxonomy of 51-based non-hybrid approaches for secure routing in MANETs is depicted in Figure 5.16.

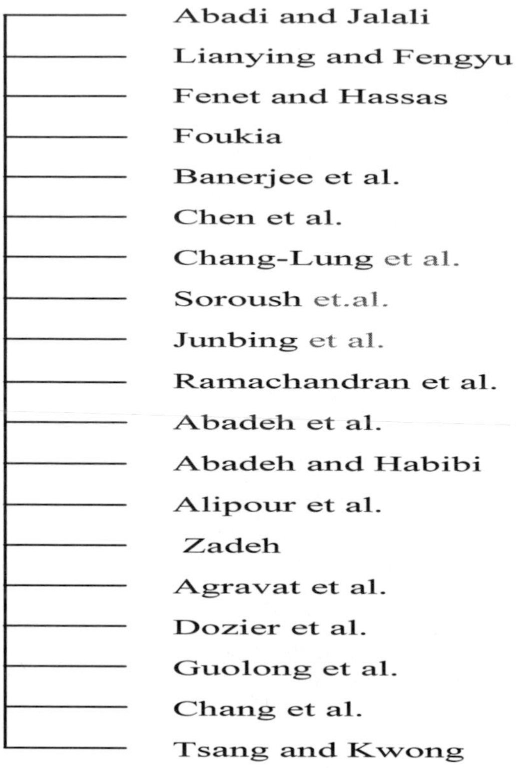

FIGURE 5.16
SI-Based Non-Hybrid Approaches for Secure Routing

5.6.2 SI-Based Hybrid Approaches

Since most PSO-based IDS are hybrid anomaly detection systems that combine methods previously discussed, it is possible to categorize them according to the existing work as follows.

Hybrid PSO and Neural Network Approaches

Michailidis et al. [45] devised an innovative method for merging PSO and artificial neural networks (ANNs) to improve intrusion detection. In their system, the PSO is executed recursively to train the network. Each particle corresponds to a synaptic weight on the network and the ANN considers the optimal synaptic weights as inputs. The input layer is built based on network connection attributes. The output layer represents normal and abnormal types.

Zhang and Benveniste [73] describe a feed-forward ANN based on wavelet analysis. The model uses a wavelet function on the hidden layer. This design achieves faster learning and avoids creation of local minima. The incorporation of such features in neural networks improves intrusion detection.

Orr [43] proposed a radial basis function (RBF) design that can be categorized as a probabilistic neural network. The classification involves a simple computation that considers the distances of the centers of neurons from the input. This scheme is included frequently in IDS design. The parameters such as distances and variances are selected manually and fed to the RBF. The selection of non-optimal parameters adversely affects the classification results.

Tian and Liu [62] used a similar method to build a hybrid PSO-ANN system. They used a mutation algorithm to help safeguard PSOs that may be entangled in local minima. This facilitates recognition of anomalies in a population and increasing the scale of a search.

Hybrid PSO and Support Vector Machine (SVM)Approaches

Wang et al. [67] proposed a real-time intrusion detection system based on PSO and SVM. Their work was the first to combine these two techniques. They used the standard particle swarm optimization (SPSO) and binary particle swarm optimization (BPSO) to determine optimal SVMparameters and extract a feature subset.

Each particle corresponds to a solution that describes the features and parameters that must be retained. The training data set, selected features, and selected parameter values are supplied as inputs to the SVM classifier that categorizes network behaviors as intrusive or normal.

Similarly, Maet al. [44] proposed an intrusion detection method combining BPSO and SVM to represent dataset features and crucial SVM parameters by particle positions. The SVM-based classification process is claimed to be precise. Several other articles in the literature concern work on hybrid PSO-SVM systems [30] [31] [41] [59] [61] [76].

Hybrid PSOand K-Means Approaches

Xiao et al. [68] proposed the K-means algorithm based on PSO for intrusion detection. They developed an IDS that combined the simplicity and good local search capabilities of the K-means algorithm with PSO. The position of each particle in the K-means algorithm is represented as a set of D dimensional centroids. A fitness function defined for each particle computes its position. If necessary, the algorithm evaluates the Pbest and G best values and updates their velocities and positions. The intent ofthe K-means algorithm is to optimize the generation of new particles via convergence to local optima at high speed and low probability. Li et al. [39] proposed a similar clustering anomaly detection algorithm based on PSO.

Hybrid Ant Colony Clustering (ACC) and Self Organizing Map (SOM)Approaches

Feng et al. [26] used a technique similar to the LF model to detect anomalies.

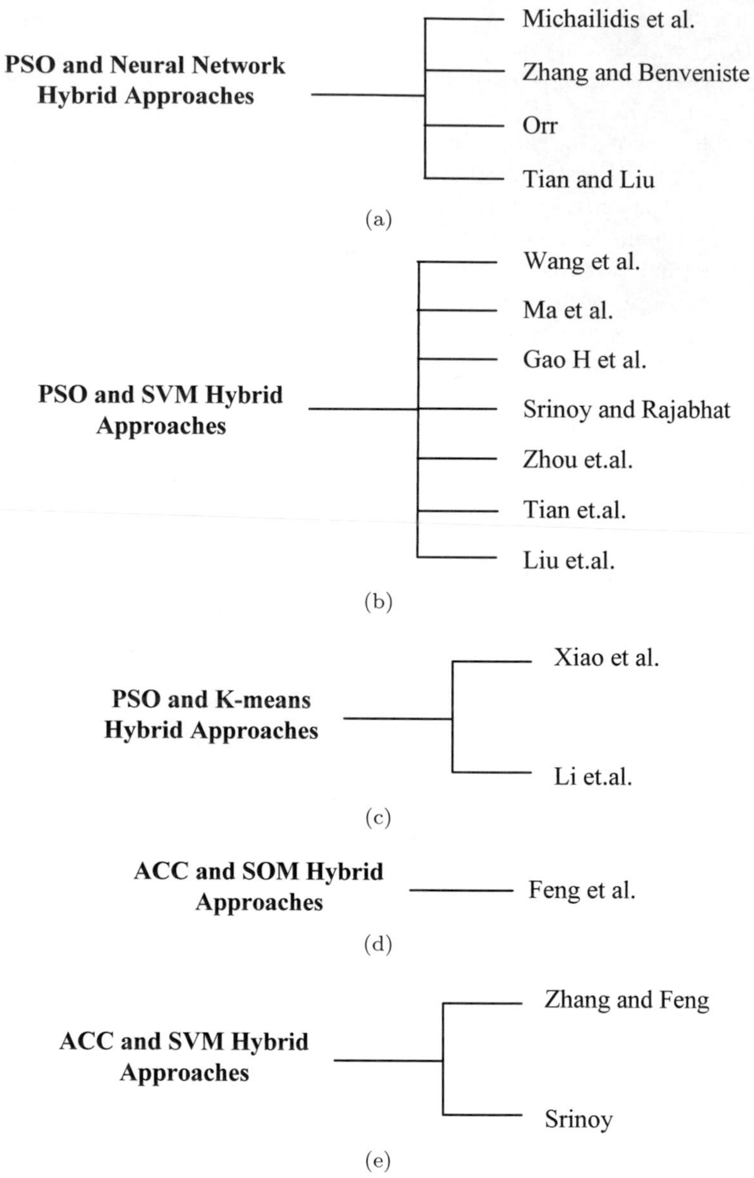

FIGURE 5.17
SI-based Hybrid Approaches for Secure Routing for MANETs

A circular area around each ant is considered a neighborhood in which pick-up and drop probabilities are calculated based on non-linear functions. The scheme helps solve linear inseparable problems. It runs a clustering phase in which formatted clusters are labeled and uses Bayes' theorem to ensure precision of the detection procedure. The same authors [27] [28] integrated their earlier method with a variant ofthe SOM[60] neural network model. A dynamic self-organizing map (DSOM) function was added to increase the efficiency of cluster formation before the ant colony clustering step proceeds.

Hybrid ACC and SVM Approaches
To increase IDS performance, Zhang and Fent [29] presented a hybrid framework combining SVM [74] and ACC for detection of network intrusions. The SVM used for clustering represents network data as data points in a multidimensional space. The authors treat the ACC algorithm as a technique for selecting data points to be used later for training at each step.

Other works on intrusion detection models combining SVM and ACC appear in the literature. Srinoy [54] devised an adaptive IDS model based on swarm intelligence and SVM. Initially, the SVM produces clusters, then the ACC refines the clusters. This is in contract with other methodologies that use ACC to reduce the amount of training required for the SVM. The results prove that the use of ACC leads to an IDS design that provides competitive DR and FAR.

Figure. 5.17 lists the taxonomies of all the hybrid approaches discussed under this section.

5.7 Summary

This concludes the detailed descriptions of various types of attacks, security issues, secure schemes and routing, and SI security solutions for MANETs.

Exercises

Part A Questions

1. To which type of attack do the message replays and DOS attacks belong in relation to security vulnerabilities?
   (a) Internal
   (b) External
   (c) Active
   (d) Passive
2. Which of the following attacks is a passive attack on security vulnerabilities?

(a) Message modification
(b) Eavesdropping
(c) Traffic Analysis
(d) Both (b) and (c)

3. Attacks carried out by nodes or groups of nodes that do not belong to a network are

(a) Internal
(b) External
(c) Active
(d) Passive

4. Attacks carried out by nodes or groups of nodes that are parts of a network are

(a) Active
(b) Passive
(c) Internal
(d) External

5. An attack that takes place in the transport layers of MANETs protocol stacks is

(a) Session hijacking
(b) Eavesdropping
(c) Rushing
(d) Traffic analysis

6. Jammers that alternate between sleeping and jamming a channel are

(a) Constant jammers
(b) Deceptive jammers
(c) Random jammers
(d) Reactive jammer

7. Jammers that are active when they sense a transmission

(a) Constant jammers
(b) Deceptive jammers
(c) Random jammers
(d) Reactive jammer

8. Nodes that disrupt the operations of any routing protocol are

(a) Selfish nodes
(b) Corrupted nodes
(c) Malicious nodes

(d) None of the above

9. An attack in which an attacker places itself in the path between the source and destination and is thus able to absorb and control network traffic flow is a

    (a) Routing attack
    (b) Blackhole attack
    (c) Rushing attack
    (d) None of the above

10. An attack in which a compromised or malicious node poses as another node by broadcasting wrong routing information and thus receiving all network traffic is a

    (a) Wormhole attack
    (b) Blackhole attack
    (c) Sinkhole attack
    (d) Replay attack

Part B Questions

1. List the different types of attacks in MANETs.
2. Draw a diagram of a protocol stack in MANETsand classify all existing security threats specific to all the layers.
3. Explain the steps involved in a routing attack and lists the changes in a routing table made by a malicious node.
4. Use an example to trace the sequence number in an AODVprotocol when a sinkhole attack is performed.
5. List the entries in the routing tables of nodes affected when a replay attack occurs.
6. Discuss various non-Sl-based MANETattack prevention approaches.
7. Discuss Sl-based hybrid approaches for secure routing for MANETs

Part A Answers

1. c
2. d
3. b
4. a
5. a
6. c

7. d
8. c
9. a
10. c

References

[1] R Hagelstrom A. A Ramanujam, J Bonney, and K Thurber. Techniques for intrusion-resistant ad hoc routing algorithms (TIARA). In *Proceedings of MILCOM Conference, Los Angeles*, pages 660–664, 2000.

[2] Habibi J and Abadeh MS. A hybridization of evolutionary fuzzy systems and ant colony optimization for intrusion detection. *ISC International Journal of Information Security*, 2(1):33–46, 2010.

[3] Soroush E, Abadeh MS, and Habibi J. Induction of fuzzy classification systems via evolutionary ACO-based algorithms. *International Journal of Simulation Systems, Science, Technology*, 9(3), 2008.

[4] Jalali S and Abadi M. An ant colony optimization algorithm for network vulnerability analysis. *Iranian Journal for Electrical and Electronic Engineering*, 2006.

[5] PB Swadas, D Agravat, and U Vaishnav. Modified ant miner for intrusion detection. In *In Proceedings of the Second International Conference on Machine Learning and Computing*, pages 228–232, 2010.

[6] P Albers and O Camp. Security in ad hoc networks: a general intrusion detection architecture enhancing trust-based approach. In *Proceedings of First International Workshop on Wireless Information Systems*, pages 1–12, 2002.

[7] M Esmaeili, M Nourhossein, H, Alipour, and E Khosrowshahi. ACOFCR: applying ACO-based algorithms to induct fcr. In *In Proceedings of World Congress on Engineering*, pages 12–17, 2008.

[8] G Avoine and S Vaudenay. Cryptography with guardian angels: bringing civilization to pirates. In *ACM Mobile Computing and Communications Review*, pages 74–94.

[9] Q Shi, B Askwith, M Merabti, and K Whiteley. Achieving user privacy in mobile networks. In *Proceedings of 13th Annual Computer Security Applications Conference*, pages 108–116, 1997.

[10] C Nita-Rotaru, B Awerbuch, D Holmer, and H Rubens. An on-demand secure routing protocol resilient to Byzantine failures. In *Proceedings of ACM Workshop on Wireless Security*, pages 21–30, 2002.

[11] S B Kamisetty, B K Bhargava, and S K Madria. Fault-tolerant authentication in mobile computing. In *Proceedings of International Conference on Internet Computing*, pages 395–402, 2000.

[12] K Wu, B Sun, and U W Pooch. Integration of mobility and intrusion detection for wireless ad hoc networks. *International Journal of Communication Systems*.

[13] A Abraham, S Banerjee, and C Grosan. Ideas: intrusion detection based on emotional ants for sensors. In *Proceedings of 5th International Conference on Intelligent Systems Design and Applications*, pages 344–349, 2005.

[14] A Abraham, PK Mahanti, S Banerjee, and C Grosan. Intrusion detection in sensor networks using emotional ants. *International Journal of Applied Science and Computations*, 12(3):152–173, 2005.

[15] J Brinkley and W Trost. Authenticated ad hoc routing at the link layer for mobile systems. 7(2):139–145, 2001.

[16] T G Brutch and P C Brutch. Mutual authentication, confidentiality and key management (Mackman) system for mobile computing and wireless communication. In *Proceedings of 14th Annual Computer Security Applications Conference*, pages 308–317, 1998.

[17] S Buchegger and J Boudec. Performance analysis of the CONFIDANT protocol: cooperation of nodes fairness in distributed ad hoc networks. In *Proceedings of MobiHoc Conference*, pages 226–236, 2002.

[18] L Buttyn and J P Hubaux. Stimulating cooperation in self-organizing mobile ad hoc networks. *ACM Journal for Mobile Networks (MONET)*.

[19] H Chin-Chuan, T Chang-Lung, and T Chun-Chi. Intrusive behavior analysis based on honey pot tracking and ant algorithm analysis. In *In Proceedings of the 43rd Annual International Carnahan*, 2009.

[20] H Chin-Chuan, T Chang-Lung, and T Chun-Chi. Intrusive behavior analysis based on honey pot tracking and ant algorithm analysis. In *In: Proceedings of the 43rd Annual 2009 International Carnahan Conference on Security Technology*, pages 248–252, 2009.

[21] L Chen and W B Heinzelman. A survey of routing protocols that support QoS in mobile ad hoc networks. *Networks*, 21(6):10–16, 2007.

[22] V Vapnik and C Cortes. Support vector networks. *Machine Learning*, 20:273–97, 1995.

[23] H Hou, J Hurley, G Dozier, and D Brown. Vulnerability analysis of immunity-based intrusion detection systems using genetic and evolutionary hackers. *Applied Soft Computing*, 7(2), 2007.

[24] J Hurley, K Cain, G Dozier, and D Brown. Vulnerability analysis of AIS based intrusion detection systems via genetic and particle swarm red teams. In *Proceedings of Congress on Evolutionary Computation*, pages 111–116, 2004.

[25] S Hassas and S Fenet. A distributed intrusion detection and response system based on mobile autonomous agents using social insects communication paradigm. In *Proceedings of First International Workshop on Security of Mobile Multiagent Systems*, pages 41–58, 2001.

[26] KG Wu, ZY Xiong, Y Zhou, Y Feng, and ZF Wu. An unsupervised anomaly intrusion detection algorithm based on swarm intelligence. In *the Proceedings of Fourth International Conference on Machine Learning and Cybernetics*, pages 3965–3969, 2005.

[27] Z Xiong, CY Ye, KG Wu, Y, Feng, and J Zhong. Intrusion detection classifier based on dynamic SOM and swarm intelligence clustering. In *Advances in Congnitive Neurodynamics*, pages 969–974, 2007.

[28] CY Ye, ZF Wu, Y Feng, and J Zhong. Clustering based on self organizing ant colony networks with application to intrusion detection. In *Proceedings of the Sixth International Conference on Intelligent Systems Design and Applications (ISDA 06)*, pages 1077–1080, 2006.

[29] Z-Y Xiong, C-X Ye, K-G Wu, Y Feng, and Z Zhong. Network anomaly detection based on DSOM and ACO clustering. *Advances in Neural Networks*, pages 947–955, 2007.

[30] X Wang, H Gao, and H Yang. Selection and detection of network intrusion feature based on BPSO-SVM. In Technical Report. Shanghai: College of Information Science and Engineering, East China University of Science and Technology, 2006.

[31] H Yang, H Gao, and X Wang. Swarm intelligence and SVM based network intrusion feature selection and detection. In Technical Report. Shanghai: College of Information Science and Engineering, East China University of Science and Technology, 2005.

[32] G Wenzhong, C Guolong, and C Qingliang. A PSO-based approach to rule learning in network intrusion detection. *Fuzzy Information and Engineering*, pages 666–673, 2007.

[33] X Meng H Yang and S Lu. Self-organized network layer security in mobile ad hoc networks. In *Proceedings of ACM MOBICOM Wireless Security Workshop*, pages 11–20, 2002.

[34] Y Huang and W Lee. A cooperative intrusion detection system for ad hoc networks,. In *Proceedings of ACM Workshop on Security of Ad Hoc and Sensor Networks*, pages 135–147, 2003.

[35] K Xu, D Gu, M Gerla, J et al. Adaptive security for multi-layer ad hoc networks. 2(5):533–547, 2002.

[36] C Chuan, H Junbing, and L Dongyang. An improved ant-based classifier for intrusion detection. In *Proceedings of Third International Conference on Natural Computation*, pages 819–823, 2007.

[37] O Kachirski and R Guha. Effective intrusion detection using multiple sensors in wireless ad hoc networks. In *Proceedings of 36th International Conference on System Sciences*, pages 57–64, 2003.

[38] LA Zadeh. Fuzzy sets. *Information Control*, 8:338–353, 1965.

[39] J Xu, B Zhao, Y Li, and G Yang. Anomaly detection for clustering algorithm based on particle swarm optimization. *Journal of Jiangsu University of Science and Technology*, 2009.

[40] L Fengyu and Z Lianying. A Swarm-Intelligence based intrusion detection technique. *International Journal of Computer Science and Network Security*, 6(7):146–150, 2006.

[41] S Liu, H Liu, and Y Jian. A new intelligent intrusion detection method based on attribute reduction and parameters optimization of SVM. In *Proceedings of the Second International Workshop on Education Technology and Computer Science*, pages 202–205, 2010.

[42] H Luo and S Lu. Ubiquitous and robust authentication services for ad hoc wireless networks. In Technical Report, Department of Computer Science, 2000.

[43] M Orr. Introduction to radial basis function networks. Technical report. Institute for Adaptive and Neural Computation Edinburgh: Edinburgh University, 1996.

[44] S Liu, J Ma, and X Liu. A new intrusion detection method based on BPSO-SVM. In *Proceedings of the International Symposium on Computational Intelligence and Design*, pages 473–477, 2008.

[45] E Georgopoulos, E Michailidis, and SK Katsikas. Intrusion detection using evolutionary neural networks. In *Proceedings of Panhellenic Conference on Informatics*, pages 8–12, 2008.

[46] P Michiardi and R Molva. CORE: a collaborative reputation mechanism to enforce node cooperation in mobile ad hoc networks. In *Proceedings of Communication and Multimedia Security Conference*, pages 107–121, 2002.

[47] N Foukia. IDREAM: intrusion detection and response executed with agent mobility. In *Proceedings of International Conference on Autonomous Agents and Multi-Agent Systems*, pages 264–270, 2005.

[48] P Papadimitratos and Z Haas. Secure routing for mobile ad hoc networks. In *Proceedings of SCS Communication Networks and Distributed Systems Modeling and Simulation Conference*, pages 27–31, 2002.

[49] AA Freitas, RS Parpinelli, HS Lopes. Data mining with an ant colony optimization algorithm. *IEEE Transactions on Evolutionary Computation*, 6(4), 2002.

[50] M Waldvogel, R Janakiraman, and Q Zhang. Indra: A peer-to-peer approach to network intrusion detection and prevention. In *Proceedings of 12th IEEE International Workshops*, pages 226–231, 2003.

[51] L Me, O Camp, R Puttini, J Percher, and R De Souza. A modular architecture for distributed IDS in MANET structures. In *Lecture Notes in Computer Science*, volume 2669, pages 91–113, 2003.

[52] MS Obaidat, C Ramachandran, and S Misra. FORK: a novel two-pronged strategy for an agent-based intrusion detection scheme in ad-hoc networks. *Computer Communications*, 31(16):3855–3869, 2008.

[53] A Abraham and V Ramos. ANTIDS: self organized ant based clustering model for intrusion detection system. In *Proceedings of Fourth IEEE International Workshop on Soft Computing as Transdisciplinary Science and Technology*, pages 977–986, 2005.

[54] S Srinoy. An adaptive IDS model based on swarm intelligence and support vector machine. In *Proceedings of International Symposium on Communications and Information Technologies*, pages 584–589, 2006.

[55] L Buttyan, S Capkun, and J P Hubaux. Self organized public-key management for mobile ad hoc networks. 2(1):52–64, 2003.

[56] K Lai, S Marti, T J Giuli, and M Baker. Mitigating routing misbehavior in mobile ad hoc networks. In *Proceedings of 6th Annual Conference on Mobile Computing and Networking*, pages 255–265, 2000.

[57] JA Habibi, E Soroush, and M SanieeAbadeh. Boosting ant-colony optimization algorithm for computer intrusion detection. In *Proceedings of IEEE 20th International Symposium on Frontiers in Networking with Applications*, 2006.

[58] R Sowjanya and H Shah. Neighborhood watch: An intrusion detection and response protocol for mobile ad hoc networks. *UMBC Technical Report*, 2002.

[59] S Rajabhat and S Srinoy. Intelligence system approach for computer network security. In *Proceedings of Fourth IASTED Asian Conference on Communication Systems and Networks*, pages 89–95, 2007.

[60] T Kohonen. Self-organizing maps. *Berlin: Springer.*

[61] J Liu and W Tian. Intrusion detection quantitative analysis with support vector regression and particle swarm optimization algorithm. In *Proceedings of International Conference on Wireless Networks and Information Systems*, pages 133–136, 2009.

[62] J Liu and W Tian. A new network intrusion detection identification model research. In *Proceedings of 2nd International Asia Conference on Informatics in Control, Automation and Robotics*, pages 9–12, 2010.

[63] S Kwong and CH Tsang. Multi-agent intrusion detection system in industrial network using ant colony clustering approach and unsupervised feature extraction. In *Proceedings of IEEE International Conference on Industrial Technology*, pages 51–56, 2005.

[64] S Kwong and W Tsang. Unsupervised anomaly intrusion detection using ant colony clustering model. In *Proceedings of 4th IEEE International Workshop on Soft Computing as Transdiciplinary Science and Technology*, pages 223–232, 2005.

[65] S Kwong and W Tsang. Ant colony clustering and feature extraction for anomaly intrusion detection. *Swarm Intelligence in Data Mining*, pages 101–121, 2006.

[66] C Tseng and P Balasubramanyam. A specification- based intrusion detection system for AODV. In *Proceedings of ACM Workshop on Security of Ad Hoc and Sensor Networks*, pages 125–134, 2003.

[67] RL Ren, Tk Wang, and X Hong. A real-time intrusion detection system based on PSO-SVM. In *Proceedings of the International Workshop on Information Security and Application*, pages 319–321, 2009.

[68] G Liu, L Xiao, and Z Shao. K-means algorithm based on particle swarm optimization algorithm for anomaly intrusion detection. In *Proceedings of Sixth World Congress on Intelligent Control and Automation*, pages 5854–5858, 2006.

[69] A Perrig, Y Hu, and D B Johnson. ARIADNE: a secure on-demand routing protocol for ad hoc networks. In *Proceedings of 8th Annual International Conference on Mobile Computing and Networking*, pages 12–23, 2002.

[70] D B Johnson, Y Hu, and A Perrig. SEAD: Secure efficient distance vector routing for mobile wireless ad hoc networks. In *Proceedings of 4th IEEE*

Workshop on Mobile Computing Systems and Applications, pages 3–13, 2002.

[71] W Lee, Y Zhang, and Y Huang. Intrusion detection techniques for mobile wireless networks. *Wireless Networks.*

[72] S Yi and R Kravets. Key management for heterogeneous ad hoc wireless networks. In Technical Report UIUCDCS-R-2002-2290, Department of Computer Science, University of Illinois, 2002.

[73] A Benveniste and Q Zhang. Wavelet networks. *IEEE Transactions on Neural Networks*, 3(6):889–898, 1992.

[74] W Feng and Q Zhang. Network intrusion detection by support vectors and ant colony. In *Proceedings of 2009 International Workshop on Information Security and Application*, pages 639–642, 2009.

[75] L Zhou and Zygmunt J. Haas.

[76] J Li, T Zhou, and Y Li. Research on intrusion detection of SVM based on PSO. In *Proceedings of International Conference on Machine Learning and Cybernetics*, pages 1205–1209, 2009.

[77] Steiner et al.

6

Conclusions and Future Directions

CONTENTS

6.1 Conclusions

MANETs are rapidly gaining popularity because of their distinguishing characteristics such as lack of fixed infrastructure, ease of deployment, no need for central administration, and increased connectivity. However, they suffer from routing complexity, need for path management and delays caused by multi-hop relaying. Providing the best QoS in MANETS is challenging.

Many researchers developed SI-based adaptive algorithms for solving real-world telecommunication problems in general and routing problems in particular. CLD for developing SI-based routing protocols to increase the efficiency of mobile wireless networks has not been explored adequately. This text attempts to demonstrate the analogy between SI concepts and MANETs working principles to readers.

The first part of Chapter 2 deals with behavioral studies of SI and non-Sl techniques In detail. The characteristics of ants, termites, fish, birds, and fireflies are relevant to SI protocols. Mammal bat behaviors relate to non-Sl schemes. Ants and termites dominate in the area of routing in MANETs.

Chapter 2 also compares the adoption of bat and termite behaviors in MANETs routing systems. The ACO working principles of a simple MANET topology are presented and the pheromone table contents of all nodes are explained in detail. Step-by-step discussions and illustrations of pheromone tables and updating steps are included for reader convenience.

In Chapter 3, the working principles of legacy routing protocols for MANETs are covered in detail and a detailed study of SI-based routing protocols is presented. The discussion focuses on the state of the art of each routing protocol, parameters required for optimization and pheromone updating and decay. Each section contains implementation details and comparisons of techniques. The chapter ends with the taxonomy of SI-based routing protocols for MANETs.

Qo5 routing protocols for MANETs are explained in Chapter 4. The focus is on termite-based routing protocols that use CLD approaches such as mobility aware termite and load-balanced termite algorithms for ensuring QoS.These hybrid QoS aware routing protocols utilize the fascinating behaviors of termites and bats to achieve network objectives.

Route discovery and maintenance, node structures, pheromone tables, and pheromone update and decay functions, and QoS parameters for each protocol are discussed. The chapter ends with detailed explanations of other QoS aware routing protocols for MANETs utilizing SI principles.

6.2 Future Directions

While SI and CLD are promising approaches for providing QoS for MANETs, certain aspects are still unexplored. The following issues and challenges are observed for further research suggested:

1. Cross layer design has been receiving the tremendous attention among researchers for increasing the efficiency of MANETs but has not been used in bio-inspired routing protocols.
2. Swarm intelligence-based routing protocols for MANETs focus on ACO and only a handful of routing protocols the behaviors termites could be found.
3. Existing termite-based routing protocols for MANETs concentrate on interpacket arrival times for pheromone update and decay over the links. Finding the stable and reliable nodes for the path with this approach is difficult as MANETs suffer from frequent link breakups due to the mobility of nodes.
4. Very few ant colony routing algorithms for MANETs have tried to mitigation stagnation. However, termite-based routing algorithms may be of help in solving the stagnation problem [1]. Load aware bio-inspired routing protocols for MANETs motivated several researchers to combat the stagnation issue efficiently.
5. A little effort could exploit the intelligent features of social insect and animal societies in designing hybrid bio-inspired routing protocols for MANETs [2].
6. The context awareness, load balancing, and QoS requirements of routing protocols for MANETs have gained the attention of several researchers and led to novel bio-inspired algorithms utilizing salient

aspects of cross layer design and behaviors of social insects and other animal societies.

References

[1] M Roth and S Wicker. 2004a.

[2] M Roth and S Wicker. 2004b.

[3] GS Jahanbakhsh et al. 2011.

Index